Frontiers of Chemical and Materials Engineering
ICoFCheM 2025

International Conference on Frontiers of Chemical and Materials Engineering (ICoFCheM 2025), Tuanku Syed Faizuddin Putra Library Hall, Universiti Malaysia Perlis (UniMAP), 11 September 2025, Malaysia

Editors

Teh Pei Leng[1], Lim Bee Ying[1], Ho Li Ngee[1], Hoo Peng Yong[1], Mohd Fitri bin Mohamad Wahid[1], Sri Raj Rajeswari Munusamy[1], Nur Maizatul Shima Adzali[1], Mohd Fitri Mohamad Wahid[1], Nur Hidayah Ahmad Zaidi

[1]Universiti Malaysia Perlis (UniMAP), Malaysia

Peer review statement

All papers published in this volume of "Materials Research Proceedings" have been peer reviewed. The process of peer review was initiated and overseen by the above proceedings editors. All reviews were conducted by expert referees in accordance to Materials Research Forum LLC high standards.

Published under License by **Materials Research Forum LLC**
Millersville, PA 17551, USA

Published as part of the proceedings series
Materials Research Proceedings
Volume 60 (2026)

ISSN 2474-3941 (Print)
ISSN 2474-395X (Online)

ISBN 978-1-64490-396-4 (Print)
ISBN 978-1-64490-397-1 (eBook)

This book contains information obtained from authentic and highly regarded sources. Reasonable efforts have been made to publish reliable data and information, but the author and publisher cannot assume responsibility for the validity of all materials or the consequences of their use. The authors and publishers have attempted to trace the copyright holders of all material reproduced in this publication and apologize to copyright holders if permission to publish in this form has not been obtained. If any copyright material has not been acknowledged please write and let us know so we may rectify in any future reprint.

Distributed worldwide by

Materials Research Forum LLC
105 Springdale Lane
Millersville, PA 17551
USA
https://mrforum.com

Manufactured in the United State of America
10 9 8 7 6 5 4 3 2 1

Table of Contents

Keyword Index

Preface

The International Conference on Frontiers of Chemical and Materials Engineering (ICoFCheM 2025) brings together researchers, lecturers, and students from across Malaysia to disseminate and exchange scientific knowledge. The conference serves as a dynamic platform for academics and researchers to share expertise, discuss recent developments, and present research findings spanning various disciplines within chemical and materials engineering. Through these interactions, ICoFCheM 2025 aims to foster collaboration, strengthen professional networks, and encourage meaningful dialogue that may lead to new discoveries and advancements in research and development.

ICoFCheM 2025 was held virtually on 11 September 2025 and attracted a total of 20 participants from various universities. Among these participants, 15 submitted manuscripts for consideration for Scopus-indexed publication in Materials Research Proceedings (MRP). These manuscripts were presented during the virtual conference sessions held on the same date.

The organizers would like to express their sincere appreciation to the Vice-Chancellor of Universiti Malaysia Perlis (UniMAP) for the continued support extended to this conference. We also extend our deepest gratitude to the Faculty of Chemical Engineering Technology for providing the opportunity for the Centre of Excellence for Frontier Materials Research to organize this meaningful event. Finally, we would like to acknowledge and thank all members of the ICoFCheM 2025 organizing committee from the Faculty of Chemical Engineering Technology for their dedication and tireless efforts in ensuring the successful execution of the conference.

Committees

Assoc. Prof. Dr. Khairel Rafizi Ahmad
Dr. Norzilah Abdul Halif
Dr. Zuraidawani Che Daud
En. Murizam Darus

Editorial Team
Assoc. Prof. Dr. Teh Pei Leng (Chief Editor)
Assoc. Prof. Dr. Lim Bee Ying
Assoc. Prof. Dr. Ho Li Ngee
Dr. Hoo Peng Yong
Dr. Sri Raj Rajeswari Munusamy
Dr. Nur Maizatul Shima Adzali
Dr. Mohd Fitri Mohamad Wahid
Dr. Nur Hidayah Ahmad Zaidi

Frontiers of Chemical and Materials Engineering - ICoFCheM 2025   Materials Research Forum LLC
Materials Research Proceedings 60 (2026) 1-9   https://doi.org/10.21741/9781644903971-1

Pyrolyzed Palm Kernel Shell (PKS) as Sustainable and Viable Biomass Reductant in Ferrous Metallurgy

Sri Raj Rajeswari Munusamy[1,2,a*], Nur Farhana Diyana Mohd Yunos[2,3,b],
Mohd Sobri Idris[1,2,c], Chen Yi Zhen[1], Nursyafiah Athirah Muhammad Reza[1],
Asshwen Raaj Mohan[1], Mohammad Nayazy Hairil Abd Wahab[1]

[1]Faculty of Chemical Engineering & Technology (FKTK), Universiti Malaysia Perlis, Kompleks Pusat Pengajian Jejawi 2, 02600 Arau, Perlis, Malaysia

[2]CoE Frontier Materials Research, Universiti Malaysia Perlis, 02600 Arau, Perlis, Malaysia

[3]Faculty of Mechanical Engineering & Technology, Kampus Alam UniMAP, Pauh Putra, 02600 Arau, Perlis, Malaysia

[a]rajeswari@unimap.edu.my, [b]farhanadiyana@unimap.edu.my, [c]sobri@unimap.edu.my

Keywords: Biomass, Desulfurization, Dephosphorization, Ferrous, Pyrolysis

Abstract. This work investigates the effect of temperature and particle size of palm kernel shell in the pyrolysis process and its carbon utilization in ferrous metallurgy. The biomass was reduced into 1-2 mm (coarser) and ≤63 μm (finer) size fractions while iron ore size fixed at ≤63 μm. Biomass particles were heated in a muffle furnace at 10 °C/min for 1 hour at 400 °C, 600 °C and 800 °C. Iron ore reduction used the same setup and heating profile but performed at 1000 °C. X-Ray Fluorescence (XRF), X-Ray Diffraction (XRD), Carbon, Hydrogen, Nitrogen, Sulfur (CHNS) analyzer and Scanning Electron Microscopy (SEM) techniques used for evaluation of the raw and reacted samples. CHNS analysis proved the increment of carbon content from 64.5% at 400 °C to 70.2% at 800 °C compared to 39.6% for raw biomass. Morphologically, the biocarbon changed from fibrous texture to porous structure with interconnected pores. XRD analysis showed the presence of carbon, iron oxide, coesite (silica) and hedenbergite (calcium iron silicate) phases in the pyrolyzed samples. At 600 °C, the ≤63 μm size biomass experienced greater weight loss up to 75.02% compared to 71.39% for coarser fractions. Reactions between iron ore and biocarbon reductant, resulted in formations of Fe_2O_3 (84.45%), SiO_2 (7.11%), CaO (2.86%), K_2O (2.16%) as the main reaction products and impurity phases. Other specific chemical compositions for steel such as MnO (0.70%), SO_3 (1.36%) and P_2O_5 (1.19%) which influence the properties like strength, toughness and embrittlement are in range and can further be optimized by desulfurization and dephosphorization.

Introduction

Biomass pyrolysis is a non-oxidative thermochemical process that decomposes organic materials to produce biochar, bio-oil and syngas [1]. This process involves the breakdown of hemicellulose, cellulose, and lignin with generation of non-condensable gases and hydrocarbons like CO_2, CO, H_2, CH_4, C_2H_4, solid residue (biochar), tar and mineral ash [1-3]. Generally, biomass are categorized into several categories based on biological diversity, origin and source such as wood residues, agricultural residues, animal residues, industrial biomass, and municipal solid waste [4]. Among this, palm kernel shell (PKS) and palm kernel shell ash (PKSA) residues found widespread applications in concrete reinforcement, aggregate and additives, partial replacement in cement, geopolymer composite, water purification, development of plastic polymer composites, fuel source, partial reinforcing materials in metal matrix composite (MMCs) etc.[5,6]. The aptness of biocarbon and bio-coke as fuels and reductants, especially PKS and PKSA in metallurgical

Frontiers of Chemical and Materials Engineering - ICoFCheM 2025   Materials Research Forum LLC
Materials Research Proceedings 60 (2026) 1-9   https://doi.org/10.21741/9781644903971-1

processes such as coking, sintering, iron ore reduction, blast furnace injections, BF-BOF route and electric arc furnace steelmaking also discussed by researchers [7,8]. Besides, other pyrometallurgical processes such as ferroalloys, copper, zinc, aluminum, titanium industries etc. [7-10] also studied globally.

Principally, biomass pyrolysis depends on several factors such as temperature, residence time, feedstock particle size, heating rate (slow, intermediate, fast or flash), catalyst, reactor type, types of biomass, pressure and inert gas flow rate [11,12]. Higher pyrolysis temperature often yields biochar with greater specific surface area, higher porosity, increased pH, carbon contents and ash [13]. Meanwhile, lower heating rates are considered as favorable for char formation whilst faster heating rates for optimization of bio-oil output, signifying the importance of pyrolysis conditions [2,13,14]. Wei et al. (2017) and Gopinath et al. (2021) demarcated C, H, O, N and S as the main organic elements in biomass whereas Si, Al, Ca, K, Na, etc. as the inorganic elements in its ash depending on the biomass variety. Apart from these factors, the effect of size is also significant as finer size particles typically results in a greater amount of biochar generated with higher surface area and porosity which is desirable for specific applications [17].

This study explores the pyrolysis behavior of PKS under controlled conditions and slow heating rate on its thermal decomposition and biochar yield. Investigations were focussed on the effect of temperature, particle size and chemical constituents of feedstock PKS biomass, knowing that product properties would differs with pyrolysis conditions and biomass variety and localities. It further investigates the viable application of lignocellulosic PKS-derived biochar as a diversified reductant with slagging potential in ferrous metallurgy.

Methodology

Sample preparation

Palm Kernel Shell (PKS) used in this study was obtained from local palm oil mill plantation in Perak and iron ore from Kedah, Malaysia. The raw PKS were dried in oven to eliminate moisture content. The samples were then crushed and grounded using hammer mill. Following this, the samples were sieved into two size fractions, i.e. 1-2 mm (coarser) and ≤63 μm (finer) to investigate the effect of particle size towards the carbonization and yield of solid biochar.

Pyrolysis and reduction reactions

About 10 g of PKS were weighed in alumina crucibles and pyrolyzed at 400 °C, 600 °C and 800 °C at a heating rate of 10 °C/min using laboratory muffle furnace. The carbon burn off/yields were computed for all the temperatures. The ≤63 μm size biochar particles obtained at 800 °C were mixed homogeneously with iron (≤63 μm) according to stoichiometry ratios (1C:2Fe) The triplicate samples (A, B and C) were heat treated in muffle furnace at 1000 °C for 1 hour solid residence time and 10 °C/min heating rate.

Analyses of raw samples and reduced products

The chemical composition of raw samples (iron and PKS) and reduced iron were identified using XRF technique, Model: Energy dispersive X-Ray Fluorescence (ED-XRF)- Shimadzhu/ EDX7000. CHNS analyzer was used to determine the chemical composition of raw and pyrolyzed PKS. The BS EN ISO 16948: 2015 test method was used for determination of C, H, N while Elementar Vario Micro Cube, Germany method used for S determination. The structural properties of raw and pyrolyzed PKS were analyzed by using the XRD-6000 Shimadzu X-Ray Diffraction machine at 1°/min in the 2θ range of 10° to 80° at 10 kV and 30 mA in 0.04° steps. Morphological analyses were carried out using SEM (JEOL JSM-6460LA) techniques. Surfaces of the samples were coated with a thin platinum layer using the Auto Fine Coater and examined using 10 kV and 20 kV voltage at different magnifications levels.

Frontiers of Chemical and Materials Engineering - ICoFCheM 2025   Materials Research Forum LLC
Materials Research Proceedings 60 (2026) 1-9   https://doi.org/10.21741/9781644903971-1

Results and Discussion
CHNS analysis of raw and pyrolyzed PKS

CHNS analysis in Table 1 proved that the carbon content of 39.6% in raw PKS was significantly increased up to 70.2% at 800 °C upon pyrolysis. This enrichment confirms the effectiveness of thermal treatment in concentrating carbon. Hydrogen content decreased from 4.9% in raw PKS to 1.9% at 800 °C. Meanwhile, nitrogen (0.6%) and sulfur (0.5%) contents are found to be comparatively low (<<1 wt%) in raw and pyrolyzed samples in line with Wei et al. (2017) who mentioned existence of up to 3-6 wt% sulfur in fossil-based fuels. Previous studies relate the decrease of H/C and O/C molar ratios upon pyrolysis with numerous associated process such as extensive dehydration as temperature increases and reduced hydrophilic feature of biochar. Referring to Table 1, major volatile loss from hydrogen and oxygen compounds seen at 400 °C while moderate carbonization observed at 600 °C and high carbonization at 800 °C. Lower H/C and O/C values also indicate high aromatization which shows the stability, resistance to degradation and suitability as solid-state carbon reductant in metal oxides [1,16].

Table 1: Ultimate analysis of raw and pyrolyzed PKS

Element	Raw PKS (%)	Pyrolyzed PKS		
		400 °C	600 °C	800 °C
C	39.6	64.5	62.0	70.2
H	4.9	3.4	2.2	1.9
N	0.6	0.6	0.6	0.4
S	0.5	0.5	0.5	0.5
O (by difference)	54.4	31.0	34.7	27.0

XRF analysis of raw PKS

Chemical composition analysis in Fig.1 shows CaO as the major inorganic constituent or building component of PKS followed by SiO_2 (14.13%), MgO (8.31%), NaO (5.02%), K_2O (4.50%), Fe_2O_3 (3.35%), P_2O_5 (0.77%) and TiO_2 (0.28%).

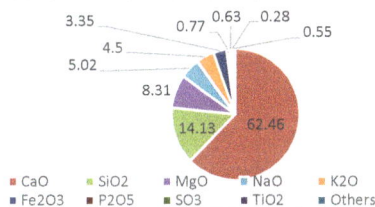

Fig. 1: XRF analysis of raw PKS

Calcium compounds are highly concentrated in PKS, attributable to natural uptake of minerals such as calcium in oil palm plants from soil. Calcium compounds accumulate in the growing oil palm plant on widespread sites, particularly in kernel shells [18]. Next highest component is SiO_2 at 14.13% which was an inert component and does not involved in the reduction reaction whereas MgO and NaO comprises the alkali and alkaline earth metal substance in the PKS [19]. Other basic oxide present in the PKS is K_2O at 4.50% while Fe_2O_3 may influence reducibility and reactivity of the material in the case of thermal treatment processes. Trace amounts of SO_3 and TiO_2 present at 0.63% and 0.28%, respectively. Presence of these oxides give insights on PKS behavior in the metallurgical

Frontiers of Chemical and Materials Engineering - ICoFCheM 2025   Materials Research Forum LLC
Materials Research Proceedings 60 (2026) 1-9   https://doi.org/10.21741/9781644903971-1

reduction reactions during thermal treatment of hematite ore, where SiO_2 and CaO can impact slag formation and reactivity [19,20].

XRD analysis of raw and pyrolyzed PKS

XRD profiles in Fig. 2 indicated prominent peaks at 200, 220, 230, 221, 202, and 260 hkl planes. In the raw PKS, the existence of silica or quartz as (coesite) was confirmed with highest crystalline peaks at about 2θ angle of 27.62° and 29.64° respectively. The presence of calcium iron silicate/hedenbergite, $CaFe(Si_2O_6)$ with a monoclinic crystal system is observed at the 221 and 202 planes. The formation of hedenbergite might be attributed by the dominant and high amount of the oxides of Ca, Si and Fe, as obtained through XRF analysis (Fig.1). As can be seen from XRF determination, CaO amounts for 62.46% whereas SiO_2 and Fe_2O_3 makes up of 14.13% and 3.35% of the total PKS constituents. Presence of iron oxide peak at 260 plane (2θ=48.657°) for the pyrolyzed PKS indicate it as the scarce element after heated if compared to raw sample. Further, the diffraction pattern also revealed carbon peak at 220 plane (2θ=26.512°) with notable increase in carbon content as temperature goes up. The emergence of carbon peaks at higher temperatures corroborates the CHNS findings (Table 1), confirming enhanced carbonization and higher carbon-rich char formation as reported by [5,25]. As the pyrolysis temperature increases from 400 °C to 800 °C, the intensity of silica (coesite) peaks in pyrolyzed samples were seen to decrease significantly if compared to raw sample, indicating removal as ash components and combinations with other oxide components. Notably, the pyrolysis process turns biomass into solid biochar rich in carbon and ash components for its inorganic equivalents. SiO_2, CaO, K_2O, Fe_2O_3, Al_2O_3, P_2O_5, etc. are among the inorganic components reported as the common chemical compositions of the palm kernel ash, depending on the feedstock biomass [5].

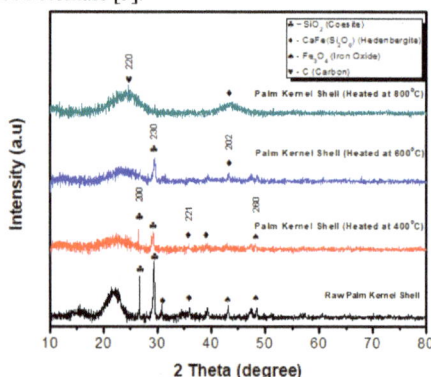

Fig. 2: XRD of raw and pyrolyzed PKS

Biochar yield (%) and post-reduction weight loss

The yield of solid biochar at three different pyrolysis temperatures (400 °C, 600 °C and 800 °C) for two particle size ranges, 1-2 mm (coarser particles) and ≤63 μm (finer particles) are produced in Fig. 3. A significant increase in biochar yield observed within the temperature range of 400 °C and 600 °C, for both size fractions. Noticeably, ≤63 μm and 1-2 mm size particles exhibited low and moderate yield, ranging from 75.02% to 75.28% and 71.39% to 75.54% as the pyrolysis temperature increased from 600 °C to 800 °C. Overall, fine particles seen to attained higher burn off and char formations compared to coarser counterparts, probably due to better heat transport and smaller temperature gradients in fine particles as described by Encinar et al., [21].

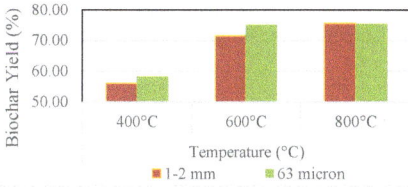

Fig. 3: Biochar yield (%) with temperature and particle size variations.

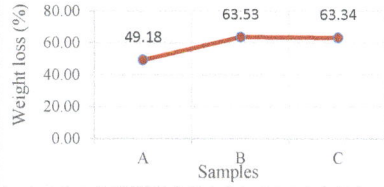

Fig. 4: Weight loss of ore-PKS biochar blends (Sampel A-C) at 1000 °C prior to reduction.

Meanwhile, the weight loss and reduction effect can be observed at 1000 °C for all the post-reacted samples of iron ore-PKS biochar blends (Samples A,B,C) as depicted in Fig.4. Slight lesser weight loss in Sample A might be due to reduced extraction of oxygen from hematite structure due to probable oxidation of sample before reduction, inhomogeneity of blending and contact between reactants for optimized reactions. Higher weight loss could be attributed by the increase of reacting moles which leads to increase of adsorption rate and, increased rate of mass transfer for diffusion and chemical reactions [22].

Chemical composition of reduced ore

XRF analysis of the raw and reduced iron are summarized in Table 2. The major oxides and impurities present in raw hematite include Fe_2O_3 (92.33%), SiO_2 (4.85%), K_2O (1.28%), P_2O_5 (0.45%), CaO (0.21%) and MnO (0.70%). Decrease in Fe_2O_3 content from 92.33% to 84.45%, confirms partial reduction as melting temperature of iron is about 1550 °C [20]. The SiO_2 addition from hematite and PKS (refer to Fig.1) participated in the iron oxide reduction and reflected by its higher post reduction amount of 7.11%.

Table 2: XRF analysis of raw and reduced ore.

Oxide %	Fe_2O_3	SiO_2	CaO	K_2O	NaO	SO_3	P_2O_5	MgO	MnO	V_2O_5	ZnO	CuO	SrO	ZrO_2	TiO_2
Hematite	92.33	4.85	0.21	1.28	-	-	0.45	-	0.70	0.07	0.06	0.03	0.01	0.01	-
Reduced Hematite	84.45	7.11	2.86	2.16	-	1.36	1.19	-	0.70	0.07	0.05	0.05	0.01	0.01	-

Higher CaO (2.86%) content was observed in reduced hematite with CaO additions primarily contributed by PKS (62.46%), resulting in higher percentage upon post-reduction. CaO presence is vital in smelting and slag formation to increase purification of reduced ore through formation of acidic or basic slag depending on the chemical balance between basic and acidic oxides in slag like CaO/SiO_2, $CaO+MgO/SiO_2+Al_2O_3$ ratios [20,23]. In this case, PKS by its own contributes flux input through substitution of CaO without external supply of lime or calcium carbonate which are the common fundamental flux material [20,23]. Increased level of SiO_2 and CaO suggest slag-forming potential of PKS in ferrous metallurgy. The occurence of SO_3 in the reduced sample indisputably is contributed from PKS as proven by CHNS analysis. Sulphur is taken by plants as sulphate and results in both organic and inorganic species in biochar [24]. Meanwhile, both the iron and PKS contributes to the presence of phosphorus in reduced products. Gao et al., [25] mentioned that a substantial amount of phosphorus which is a detrimental element in most steel groups could enters into the reduced iron from gangue minerals during the reduction process. In steelmaking process, removal of excess sulphur and phosphorus in metal phase accomplished through desulfurization and dephosphorization process using various agents [25]. Trace quantities of manganese oxide (MnO) of 0.700% and vanadium oxide (V_2O_5) at 0.067% from raw iron could improve the strength and mechanical properties of iron during reduction [26]. Chemical composition analysis also proved that other trace elements in the ore and PKS are all within the typical range of minor components in metals.

Morphological transformations in PKS and reduced ore

SEM imaging discovered significant morphological changes post-pyrolysis. Raw PKS (Figs. 5a-b) displayed irregular, flaky and elongated shaped particles with brittle structure. At higher magnification, the existence of fibrous, layered structure and small pores and pits can be seen clearly in raw PKS.

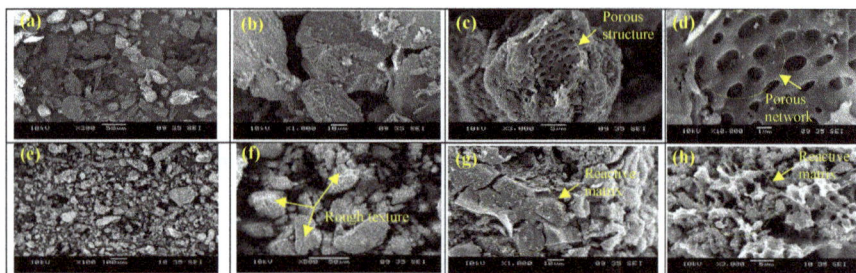

Fig.5: (a) Overall shape and size of raw PKS at 300 × Magnification, (b) Fibrous texture and small pores in raw PKS at 1000 ×, (c) Porous structure in internal walls of pyrolyzed PKS at 3000 ×, (d) Larger pores at surface of pyrolyzed PKS at 10,000 ×, (e) Overall shape and size of raw iron particles at 100 ×, (f) Solid, compact and rough texture of raw iron particles at 500 ×, (g) Reactive matrix of reduced ore-PKS biochar blends at 1000 ×, (h) Reduced matrix of ore-PKS biochar blends at 3000 ×.

Upon subjection to higher decomposition temperature and breakdown of lignocellulosic components with release of volatiles and gases [13,16], voids and porous structure were developed within the fibre network as depicted in Figs. 5(c-d). Diez et al., [26] described hemicellulose as the first lignocellulosic structure to decompose between 200-300 °C, followed by cellulose between 250-380 °C while lignin has widest decomposition range from 200 °C-1000 ° C due to its complex structure. Meanwhile, the hematite particles shown in Figs. 5(e-f) are angular to irregular in shape, compact in structure and distributed in size (≤63 μm). Smaller particles may assist the ore in being more reactive during reduction process due to larger surface area [27]. Microcracks, roughness and crevices on particles surface may allow interactions with reducing or oxidising agents by enhancing surface area available for reactions [28]. Comparatively, Figs. 5(g-h) revealed the morphology of partially reduced matrix. At 1000 °C, the ore particles seen to exhibit a loose structure with developed porous system and some bright regions are visible. Such phenomenon of iron reduction were confirmed in previous research by Man et al., [22]. Some larger and irregularly shaped particles can be seen scattered in the porous networking suggesting the formations of reduced phases, slag forming minerals and reactive matrix which is amenable for metallurgical process. According to Zulkarnia et al.[29], porous and interconnected structure of biochar as reducing agent would provide desired interaction with hematite ore for effective reduction.

Conclusion

This study demonstrates the potential of pyrolyzed PKS as a sustainable and effective biomass reductant in ferrous metallurgy. The pyrolysis process significantly enhanced the carbon content of PKS, with higher temperature yielding more porous and carbon rich biochar. These structural and compositional changes were confirmed through CHNS, XRF, XRD and SEM analyses. The biochar produced from PKS not only exhibited favorable characteristics as carbon and fuel source for metallurgical reduction but also contributes essential flux minerals such as CaO and MgO. These oxides play a critical role in impurity removal and slag formation during iron ore reduction. The presence of minor impurities such as sulfur and phosphorus was within acceptable limits and can be further managed through standard refining process. Future work will focus on reduced iron phase analysis and mettalization rates. Overall, PKS biochar presents a promising, viable and sustainable role in ferrous metallurgical system.

Acknowledgements

The authors wish to acknowlege the support from Research Management and Innovation Center (RMIC) and Faculty of Chemical Engineering & Technology (FKTK), Universiti Malaysia Perlis (UniMAP) for this research.

References

[1] F. Amalina, A.S.A. Razak, S. Krishnan, H. Sulaiman, A.W. Zularisam, & M. Nasrullah, Biochar production techniques utilizing biomass waste-derived materials and environmental applications- A review, Journal of Hazardous Materials Advances. (2022). https://doi.org/10.1016/j.hazadv.2022.100134

[2] G.R. Monga, C.T. Chongb, W.W.F. Chongc, Jo-Han Ngd, H.C. Onge, V. Ashokkumarf, Manh-Vu Trang, S. Karmakarh, B.H.H. Gohb, M.F.M. Yasini, Progress and challenges in sustainable pyrolysis technology: Reactors, feedstocks and products, Fuel. 324 (2022) 124777. https://doi.org/10.1016/j.fuel.2022.124777

[3] A. Demirbaş, Biomass resource facilities and biomass conversion processing for fuels and chemicals, Energy Conversion and Management. 42 (2001)11, 1357-1378. https://doi.org/10.1016/S0196-8904(00)00137-0

[4] L. Fernández, S.F. Pérez, J. Fernández-Ferreras, T. Llano, Microwave-Assisted Pyrolysis of Forest Biomass, Energies. 17 (2024) 4852. https://doi.org/10.3390/en17194852

[5] P.P. Ikubanni, M. Oki, A.A. Adeleke, A.A. Adediran, O.S. Adesina, Influence of temperature on the chemical compositions and microstructural changes of ash formed from palm kernel shell, Result Eng. 8 (2020), 1-9. https://doi.org/10.1016/j.rineng.2020.100173

[6] A.A. Adeleke, P.P. Ikubanni, T.A. Orhadahwe, C.T. Christopher, J.M. Akano, O.O. Agboola, S.O. Adegoke, A.O. Balogun, R.A. Ibikunle (2021). Sustainability of multifaceted usage of biomass: A Review, Heliyon. 7 (2021), e08025, 1-19. https://doi.org/10.1016/j.heliyon.2021.e08025

[7] S. Wang, Y. Chai, Y. Wang, G. Luo, A. S., Review on the Application and Development of Biochar in Ironmaking, Metals. 13 (2023), 1844. https://doi.org/10.3390/met13111844

[8] T.R. Sarker, D.Z. Ethen, S. Nandha. Decarbonization of Metallurgy and Steelmaking Industries Using Biochar: A Review. Chemical Engineering Technology, 47 (2024) (No.12), e202400217, 1-12. https://doi.org/10.1002/ceat.202400217

[9] J. Zhang, H. Fu, Y. Liu, H. Dang, L.Ye, A.N. Conejio, & R. Xu, Review on biomass metallurgy: Pretreatment technology, metallurgical mechanism and process design, International Journal of Minerals, Metallurgy and Materials. 29 (2022) (6), 1133-1149. https://doi.org/10.1007/s12613-022-2501-9

[10] R.Wei, K. Meng, H. Long, & C. Xu, Biomass metallurgy: A sustainable and green path to a carbon-neutral metallurgical industry, Renewable and Sustainable Energy Reviews, 2024. https://doi.org/10.1016/j.rser.2024.114475

[11] D. Aboelela, H. Saleh, A.M. Attia, Y. Elhenawy, T. Majozi, M. Bassyouni, Recent Advances in Biomass Pyrolysis Processes for Bioenergy Production: Optimization of Operating Conditions, Sustainability. 15 (2023), 11238, 1-30. https://doi.org/10.3390/su151411238

[12] M. Garcia-Perez, X.S. Wang, J. Shen, M.J. Rhodes, F.Tian, W.-J. Lee, H.Wu, & C.-Z. Li, Fast Pyrolysis of Oil Mallee Woody Biomass: Effect of Temperature on the Yield and Quality of Pyrolysis Products, Industrial & Engineering Chemistry Research. 47(2008) (6), 1846-1854. https://doi.org/10.1021/ie071497p

[13] A. Tomczyk, Z. Sokolowska, P. Boguta, Biochar physicochemical properties: Pyrolysis temperature and feedstock kind effects, Reviews in Environmental Science and Bio/Technology. 19(2020) (1), 191-215. https://doi.org/10.1007/s11157-020-09523-3

[14] D. Chiaramonti, A. Oasmaa, & Y. Solantausta, Power generation using fast pyrolysis liquids from biomass, Renewable and Sustainable Energy Reviews. 11(2007) (6), 1056-1086. https://doi.org/10.1016/j.rser.2005.07.008

[15] R.Wei, L. Zhang, D. Cang, J. Li, X. Li, C.C. Xu, Current status and potential of biomass utilization in ferrous metallurgical industry, Renewable and Sustainable Energy Reviews. 68 (2017), Part 1, 511-524. https://doi.org/10.1016/j.rser.2016.10.013

[16] A. Gopinath, G. Divyapriya, V. Srivastava, A.R. Laiju, P.V. Nidheesh, Conversion of sewage sludge into biochar: A potential resource in water and wastewater treatment, Environmental Research. 194 (2021) 110656. https://doi.org/10.1016/j.envres.2020.110656

[17] J. O. Ighaloa, J. Conradiec, C. R. Ohorod, J. F. Amakue, K. O. Oyedotunf, N. W. Maxakatog, K. G. Akpomiec, E. S. Okekei, C. Olisahl, A. Malloumc, K. A. Adegokeg, Biochar from coconut residues: An overview of production, properties, and applications, Industrial Crops & Products 204 (2023), 117300. https://doi.org/10.1016/j.indcrop.2023.117300

[18] F. Abnisa, A. Arami-Niya, W.M.A. Wan Daud, J.N. Sahu, & I.M. Noor, Utilization of oil palm residues to produce bio-oil and bio-char via pyrolysis, Energy Conservation and Management. 76 (2013), 1073-1082. https://doi.org/10.1016/j.enconman.2013.08.038

[19] R. K. Liew, W. L. Nam, M. Y. Chong, X. Y. Phang, M. H. Su, P. N. Y. Yek, N. L. Ma, C. K. Cheng, C. T. Chong, & S. S. Lam, Oil palm waste: An abundant and promising feedstock for microwave pyrolysis conversion into good quality biochar with potential multi-applications, Biowaste for Energy recovery and Environmental Remediation,. 115 (2018), 57-69. https://doi.org/10.1016/j.psep.2017.10.005

[20] S. Biswal, F. Pahlevani, W.Wang, V. Sahajwalla, Reduction Behavior of Hematite-Biowaste Composite Pellets at Melting Temperature, Steel Research International. 95 (2024), 2300454, 1-11. https://doi.org/10.1002/srin.202300454

[21] J.M. Encinar, J.F. Gonzalez, J. Gonzalez, Fixed-bed pyrolysis of Cynara cardunculus L. Product yields and compositions, Fuel Processing Technology. 68 (2000), 209-222. https://doi.org/10.1016/S0378-3820(00)00125-9

[22] Y. Man, J.X. Feng, Y.M. Chen, J.Z. Zhou, Weight loss and direct reduction characteristics of iron ore-coal composite pellets, Journal of Iron and Steel Research International. Vol. 21 (2014) (12), 1090-1094. https://doi.org/10.1016/S1006-706X(14)60188-6

[23] T. Umadevi, P. Kumar, N.F. Lobo, M. Prabu, P.C. Mahapatra, M. Ranjan, Influence of Pellet Basicity (CaO/SiO2) on Iron Ore Pellet Properties and Microstructure, ISIJ International. Vol. 51(2011) (1), 14-20. https://doi.org/10.2355/isijinternational.51.14

[24]Kong, S.H., Loh, S.K., Bachmann, R.T., Zainal, H., Cheong, K.Y. (2019). Palm Kernel Shell Biochar Production, Characteristics and Carbon Sequestration Potentiol. Journal of Oil Palm Research, 31 (3), 508-520.

[25] P. Gao, G.-F. Li, Y.-X. Han, & Y.-S. Sun, Reaction Behavior of Phosphorus in Coal-Based Reduction of an Oolitic Hematite Ore and Pre-Dephosphorization of Reduced Iron, Metals. 6(2016)(4), 82. https://doi.org/10.3390/met6040082

[26] D. Diez, A.Urueña, R. Piñero, A. Barrio & T.Tamminen, Determination of Hemicellulose, Cellulose and Lignin Content in Different Type of Biomasses by Thermogravimetric Analysis and Pseudocomponent Kinetic Model (TGA-PKM Method), Processes. 8 (2020), 1048. doi:10.3390/pr8091048. https://doi.org/10.3390/pr8091048

[27] Y. Qu, Y. Yang, Z. Zou, C. Zeilstra, K. Meijer, & R. Boom, Reduction Kinetics of Fine Hematite Ore Particles with a High Temperature Drop Tube Furnace, ISIJ International. 55 (2015) (5), 952-960. https://doi.org/10.2355/isijinternational.55.952

[28] S. K. Jena, H. Sahoo, S. S. Rath, D. S. Rao, S. K. Das, & B. Das, (2015). Characterization and Processing of Iron Ore Slimes for Recovery of Iron Values, Mineral Processing and Extractive Metallurgy Review. 36 (2015) (3), 174-182. https://doi.org/10.1080/08827508.2014.898300

[29] A. Zulkania, R. Rochmadi, M. Hidayat, & R. B. Cahyono, Reduction Reactivity of Low Grade Iron Ore-Biomass Pellets for a Sustainable Ironmaking Process. Energies.15 (2022) (1). https://doi.org/10.3390/en15010137

Frontiers of Chemical and Materials Engineering - ICoFCheM 2025   Materials Research Forum LLC
Materials Research Proceedings 60 (2026) 10-19   https://doi.org/10.21741/9781644903971-2

Fabrication of Polylactic Acid Membrane Assisted with Zinc Oxide for Methyl Orange Dye Removal

NURUL Izzah[1], AMIRA M Nasib[1,a] *, MOHAMAD SYAHMIE bin Mohamad Rasidi[1]
and MOHAMMAD FIRDAUS bin Abu Hasim[2]

[1]Faculty of Chemical Engineering & Technology, Kompleks Pusat Pengajian Jejawi 3, Kawasan Perindustrian Jejawi, Universiti Malaysia Perlis (UniMAP), Arau, Perlis, Malaysia

[2]Faculty of Mechanical Engineering & Technology, Universiti Malaysia Perlis, Kampus Tetap Pauh Putra, Perlis, Malaysia

[a]Centre of Excellence for Frontier Materials Research, Taman Pertiwi Indah, Jalan Kangar - Alor Setar, Kampung Seriap, 01000 Kangar, Perlis

*amira@unimap.edu.my

Keywords: Polylactic Acid, Polymeric Membranes, Membrane Filtration, Photocatalyst

Abstract Conventional membranes show limited efficiency in removing persistent dyes like methyl orange, highlighting the need for modified membranes with improved hydrophilicity and performance. Asymmetric PLA membranes incorporated with ZnO nanoparticles were fabricated via non-solvent induced phase separation (NIPS) to investigate the effects of ZnO loading on their physical and performance properties. Polymer solutions containing 0–0.75 wt.% ZnO were cast and immersed in a 60:40 methanol–water bath at room temperature. The membranes were characterized for morphology, hydrophilicity, porosity, flux, permeability, and methyl orange (MO) dye removal. Increasing ZnO content improved hydrophilicity (contact angle 79.6° to 57.5°) and performance up to 0.25 wt.%, achieving the highest flux and dye removal (60.9 L/m²·h and 68.8%, respectively). Beyond this level, agglomeration reduced porosity and performance. Photocatalytic degradation further enhanced MO dye removal to 71.9%. Overall, 0.25 wt.% ZnO yielded optimal structure and efficiency for sustainable dye wastewater treatment. Higher loadings led to agglomeration and reduced performance, while photocatalytic activity further enhanced methyl orange removal efficiency.

Introduction

Water pollution has become a major environmental issue due to the discharge of untreated industrial wastewater, which threatens ecosystems and human health. Industrial effluents containing synthetic dyes and heavy metals are particularly hazardous because of their stability, toxicity, and persistence. Common treatment methods include coagulation, adsorption, biological degradation, and membrane filtration [1]. Among these, membrane filtration is preferred for its high efficiency (BOD 70–90%, COD 60–85%, TSS >90%), simplicity, and ability to remove dissolved dyes and suspended solids without harmful by-products [2,4]. In this study, a microfiltration membrane was chosen for its suitable pore size and enhanced dye removal through ZnO nanoparticle modification. However, membrane fouling remains a major limitation, reducing water flux and purification efficiency [3,4].

Alosaimi et al. [5] modified sulfonated polyethersulfone (sPES) ultrafiltration membranes with polysulfopropyl acrylate (PSPA)-coated ZnO nanoparticles to enhance antifouling performance. This modification improved the membranes' hydrophilicity, reducing the water contact angle from 79 to 55° and increasing water flux by 254% compared to unmodified PES membranes. In Mezher et al. (2024), silane-functionalized ZnO nanoparticles (0–2 wt %) were incorporated into polyethersulfone (PES) membranes to treat concentrated textile wastewater via nanofiltration. The

Frontiers of Chemical and Materials Engineering - ICoFCheM 2025   Materials Research Forum LLC
Materials Research Proceedings 60 (2026) 10-19   https://doi.org/10.21741/9781644903971-2

ZnO modification led to notable changes in surface morphology and improved hydrophilicity, reducing the water contact angle from 6.5° to 47.9° at higher loadings. Despite a modest increase in average pore size (from 2.3 to 3.98 nm), the modified membranes maintained high separation ability while achieving nearly double the pure water flux (from 25 to 53 $L \cdot m^{-2} \cdot h$) at optimal ZnO content. The study also reports enhanced antifouling behavior and long-term stability, making PES/ZnO nanocomposite membranes a promising option for dye-laden textile effluent treatment [6].

Polylactic acid (PLA), a biodegradable thermoplastic polyester derived from renewable resources, has emerged as an environmentally sustainable alternative for membrane fabrication. [8]. The integration of ZnO as a photocatalyst in PLA membranes has been explored to improve membrane hydrophilicity, reduce fouling, and enhance dye removal efficiency through photocatalytic degradation [9]. In previous research, PLA/ZnO composite membranes were successfully fabricated via electrospinning with ZnO loadings of up to 40 wt.% [10]. Incorporation of 5–10 wt.% ZnO resulted in improved mechanical properties, while further increases (≥20 wt.%) led to structural deterioration. Enhanced antibacterial activity against *S. aureus* and *E. coli* was observed in ZnO-rich membranes, which was attributed to the greater exposure of nanoparticles on the membrane surface. Meanwhile, Hir et al. researched on polyethersulfone (PES)/ZnO hybrid film photocatalysts for MO degradation shows that photocatalytic efficiency increases with ZnO content up to an optimal level [11]. The PES/ZnO film with 17 wt.% ZnO achieves the highest methyl orange degradation rate of 98% due to more surface active sites that enhance UV absorption.

The PLA membrane was fabricated through the incorporation of ZnO via the phase inversion technique, a well-established method for producing porous polymer membranes. This technique mimics a demixing process, enabling the controlled transition of a polymer solution from a liquid state to a solid phase [11]. Thus, in this study, PLA membranes were fabricated by incorporating ZnO nanoparticles using the phase inversion method. The work focuses on assessing how different ZnO concentrations influence the membranes' hydrophilicity, porosity, and water contact angle. Membrane performance was further evaluated in terms of water flux, permeability, and dye rejection efficiency.

Experimental

Material and Chemicals:

Polylactic acid (PLA) and zinc oxide (ZnO) were procured from Sigma Aldrich through Fisher Scientific. N,N-dimethylacetamide (DMAc, ≥95% purity) and Methyl Orange dye were supplied by Merck Millipore Corporation. Distilled water was used as the strong non-solvent, while methanol (≥99% purity), serving as the weak non-solvent, and iso-propanol, used as the membrane wetting liquid, were obtained from SourceL HmBG Chemicals Malaysia.

Membrane Fabrication:

A homogeneous PLA polymer solution with a concentration of 16 wt.% was prepared by dissolving PLA pellets in DMAc solvent, along with predetermined amounts of ZnO. The mixture was placed in a water bath shaker and agitated continuously at 90 rpm while maintaining the temperature at 70°C for 1 hour to ensure thorough mixing and dissolution. Degassing was carried out at room temperature until a clear solution was obtained.

The flat sheet PLA membranes were fabricated using the non-solvent-induced phase separation (NIPS) method for its simplicity and effectiveness. A homogeneous polymer solution containing various ZnO concentrations (Table 1) was cast on a clean glass plate with a 500 ± 10 μm knife gap. The film was exposed to air for 30 s to allow partial solvent evaporation, then immersed in a coagulation bath of 60 wt.% methanol and 40 wt.% distilled water for 24h to complete phase separation. The membrane was then removed and dried at room temperature for 24h.

Table 1. Composition by (wt.%) of prepared polymer solution.

Membrane Label	Polylactic Acid (PLA) (wt.%)	Zinc Oxide (ZnO) (wt.%)	Dimethylacetamide (DMAc) (wt.%)
S1	16	0.00	84.00
S2	16	0.25	83.75
S3	16	0.50	83.50
S4	16	0.75	83.25

Membrane characterization:

A scanning electron microscope (SEM-EDX) model JSM6260 LE JEOL was used for the morphological characterization of the membrane structure. To minimize electric charge interference during testing, a 10 nm gold layer was sputtered onto the samples to minimize the charging effect [12]. Cross-sectional samples were fractured using liquid nitrogen and attached to the SEM holder. The membrane microstructure was examined by SEM under high vacuum at an accelerating voltage of 10 kV, and images were recorded at magnifications of 100× and 500×.

The contact angle of the membranes was measured using the sessile drop method to assess hydrophilicity. A contact angle less than 90° indicates a hydrophilic surface [13]. A microsyringe placed a 5 µL droplet of distilled water on the membrane. A high-quality image was captured, and the droplet's contour was analyzed with Image-J software. Average contact angles were calculated from three measurements at different areas of the membrane.

The gravimetric method was used to determine the porosity of the membrane with specific sizes of 2 x 2 cm. The porosity of a membrane was tested by soaking it in isopropanol for 10 minutes and then using filter paper to dry the membrane surface. The membrane's weight was measured before and after isopropanol absorption by the membrane. The porosity of membranes was calculated using Eq. 1 [14], where W_w is the mass of the wet membrane in (g), W_d is the mass of dry membrane in ρ_w (g), and ρ_p are the density of iso-propanol and PLA membrane in (g/cm³)8

$$\epsilon\ (\%) = \frac{\frac{W_w - W_d}{\rho_w}}{\frac{W_w - W_d}{\rho_w} + \frac{W_d}{\rho_p}} \times 100 \tag{1}$$

Flux and Permeability:

Flux was determined by measuring the volume of pure water permeate against the applied pressure, calculated using Eq. 2. where J represents the flux (L/m²·h), V is the volume of permeated water (L), A is the membrane area (m²), and Δt is the operating time in hours (h). Membrane permeability was calculated using Eq. 3, where P represents the membrane permeability (L/m²·h·bar) and Δp represents the transmembrane pressure of 1.75 bar [15].

$$J = \frac{V}{A \Delta t} \tag{2}$$

$$P = \frac{J}{\Delta p} \tag{3}$$

Dye Rejection:

The UV-Vis spectrophotometer measured the concentration of MO dye after filtration at 462.10 nm [16]. Each sample was tested for absorbance three times for accuracy and removal efficiency was calculated using Eq. 4, where C_0 and C_e are the initial and the equilibrium MO dye concentrations (mg/L).

$$(\%) Removal = \frac{(c_0 - c_e)}{c_0} \times 100 \tag{4}$$

Photocatalytic Reaction:

The PLA/ZnO membranes demonstrate photocatalytic activity for degrading organic pollutants under UV irradiation from a 254nm pen-ray photochemical quartz lamp. This was confirmed by degrading a pollutant solution and composite membranes in photocatalytic reactions. The solution reached adsorption-desorption equilibrium in the dark for 30 minutes before exposing the membranes to MO dye. UV-Vis analysis determined the dye concentration and the photodegradation was calculated using Eq. 5, where c_i represent the initial concentration of methyl orange dye, and c_f denote the final concentration of the dye (mg/L) [16].

$$Photocatalytic \; degradation(\%) = \frac{(c_i - c_f)}{c_i} \times 100 \tag{5}$$

Results and Discussion

Membrane Structure:

The surface morphology and cross-section of PLA membranes incorporating varying concentrations of ZnO nanoparticles were examined utilizing Scanning Electron Microscopy (SEM) and Energy Dispersive X-ray (EDX) spectroscopy. Fig. 1 presents the surface morphology (top) and cross-sectional structure (cross) of the PLA/ZnO membranes together with the EDX spectrum, revealing the effect of ZnO incorporation on membrane structure. Fig. 1 S1(top) indicates that the pure PLA membrane contains zero ZnO exhibits a smooth and homogeneous surface with minimal roughness, suggesting a dense and predominantly non-porous structure. The incorporation of 0.25 wt.% ZnO significantly alters the surface morphology, increasing roughness and forming a textured, network-like structure. This modification enhances the filtration capacity by providing additional interaction sites while preserving structural integrity [17]. At 0.5 wt.% ZnO (Fig. 1 S3(top)), the roughness is more pronounced with larger pores, indicating enhanced porosity and fluid transport. This is consistent with findings by Ding et al. [18], who noted that higher nanoparticle concentrations increase porosity and water flux. The membrane containing 0.75 wt.% ZnO (Fig. 1 S4(top)), exhibits a highly irregular surface with large pits and noticeable nanoparticle agglomeration, leading to uneven pore distribution. While this increases porosity, it may compromise mechanical stability [19-20].

The images of membrane cross sectional area (Fig. 1(cross)) shows that all membranes display an asymmetrical structure, with a dense upper layer and a porous sublayer featuring finger-like formations essential for permeation and rejection. The morphologies of the asymmetric membranes made by phase inversion are significantly influenced by the system's thermodynamic and kinetic properties and the addition of ZnO. Pure PLA membranes have a smooth upper layer and distinct finger-like formations in the sublayer, attributed to the strong affinity between the solvent (DMAC) and non-solvent (distilled water), which speeds up the phase inversion. White spots on the surface may indicate polymer clumps from uneven solidification. When ZnO is added, the asymmetric structure, particularly the finger-like pore size, changes with increasing ZnO concentration. Without ZnO, the finger-like structures are notably broad, but higher concentrations of ZnO decrease their prevalence, as indicated by Murali et al. [22].

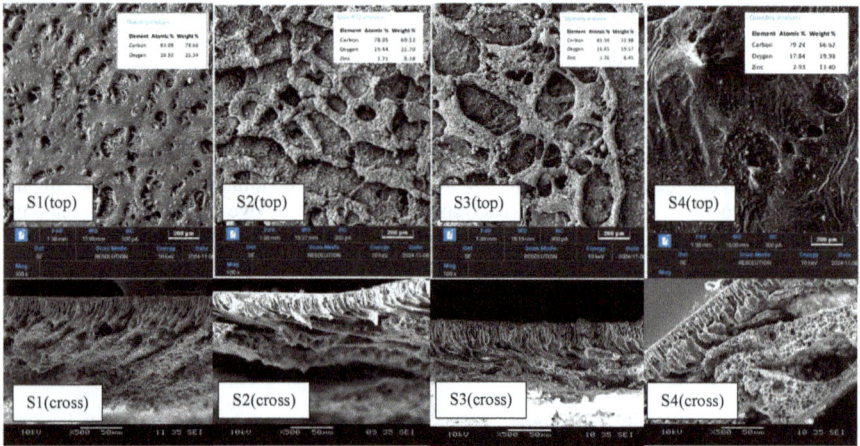

Fig. 1. Top surface morphology and cross sectional area of PLA/ZnO membrane

The EDX spectrum of pure PLA membrane shows clear peaks for carbon (C) and oxygen (O), reflecting the chemical composition of PLA. The high carbon content shows it contains many carbon-rich chains, while the oxygen comes from ester groups. There are no zinc peaks, confirming that ZnO is not present, which matches the SEM images that show a smooth and non-porous surface. Addition of 0.25 wt.% ZnO results in visible zinc peaks in the EDX analysis. This confirms the addition of ZnO nanoparticles. The increased oxygen levels suggest there are interactions between ZnO and PLA, which may create Zn-O bonds and make the surface more water-attracting. These ZnO nanoparticles disrupt the dense structure, increasing the surface area, which aligns with the surface roughness seen in the SEM images. For 0.5 wt.% ZnO, the EDX analysis shows more zinc while maintaining high carbon levels. The stable oxygen suggests a balanced chemical composition. The rise in zinc relates to greater surface roughness and a more developed porous network, which improves filtration efficiency [21]. At 0.75 wt.% ZnO, the zinc content has significantly increased, indicating a high level of ZnO and coinciding with increased roughness and pore structures. The higher amounts of oxygen and zinc suggest that ZnO is clustering. This clustering increases hydrophilicity but may also lead to agglomeration, which could affect the uniformity and mechanical properties of the membrane [20].

Water Contact Angle:

The water contact angle, a common parameter used to describe the hydrophilicity of the membrane's surface, is shown in Fig. 2(a). It was observed that the contact angle decreased as the amount of ZnO increased from 0 to 0.75 wt.%. The average water contact angle for S1 without ZnO was 74.16°, which decreased to 57.36° with the addition of ZnO up to a composition of 0.75 wt.%. This trend can be attributed to the hydrophilic nature of ZnO, leading to increased membrane hydrophilicity. The hydrophilic ZnO presence in the membrane structures resulted in lower contact angles, allowing a larger fraction of water to diffuse through the membrane. Alhoshan et al. [7] found that adding varying amounts of ZnO to PSu membranes decreased contact angle (CA) values, indicating improved hydrophilicity. The addition of ZnO nanoparticles lowered CA from 67 to 61°. Similarly, Zhao et al. [12] studied PES/ZnO membranes and found that adding ZnO nanoparticles reduced CA values, enhancing hydrophilicity; PES membranes had a CA of ~67°,

Frontiers of Chemical and Materials Engineering - ICoFCheM 2025   Materials Research Forum LLC
Materials Research Proceedings 60 (2026) 10-19   https://doi.org/10.21741/9781644903971-2

while 0.017 wt.% ZnO saw CA drop to around 46° due to the hydrophilic nature of ZnO nanoparticles.

Porosity:

The porosity of each membrane was calculated and is illustrated in Fig. 2(b). The variation in porosity is linked to the concentration of ZnO in the membrane. Initially, with 0.25 wt.% ZnO, the membrane porosity peaks at 80.42%. However, as the concentration increases to 0.75 wt.%, membrane porosity declines to 70.79%. This trend is due to the hydrophilic properties of ZnO, which enhance water exchange and initial porosity. At higher concentrations (0.5 and 0.75 wt.%), ZnO nanoparticle aggregation reduces porosity [23]. Moradihamedani et al. [24] found that PSf membranes exhibited increased porosity from 28.68 to 50.51% with ZnO rising from 0 to 1 wt.%. However, further increases to 3 and 5 wt.% decreased porosity to 17.95%, attributed to delayed phase separation affecting pore formation.

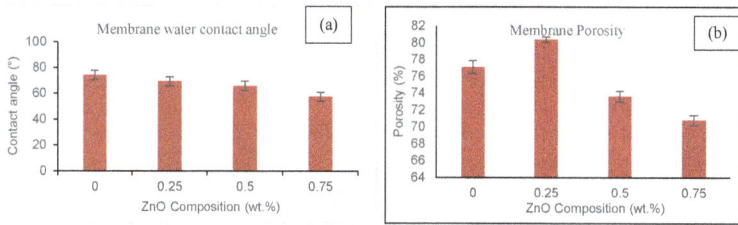

Fig. 2 Effect of ZnO Composition on PLA membrane (a) water contact angle and (b) membrane porosity

Flux and Permeability:

Fig. 3(a) shows the relationship between ZnO composition (wt.%) and pure water flux (L/m²·h) for PLA/ZnO membranes. The flux peaks at 60.88 L/m²·hr at 0.25 wt.% ZnO, attributed to improved hydrophilicity and uniform ZnO nanoparticle distribution enhancing water permeability. However, as ZnO composition increases to 0.5 wt.%, flux decreases due to nanoparticle aggregation, pore blockage, and reduced effective pore size. Similar trends are reported for other membranes. Shen et al. [23] found a 254% improvement in PES/ZnO membranes' flux, peaking at 125.40 kg/m²·h with 0.2 g ZnO, before decreasing as aggregation reduced porosity. At lower ZnO loadings, the nanoparticles are well-dispersed within the polymer matrix, providing abundant surface hydroxyl groups that enhance hydrophilicity and facilitate water transport through the membrane. However, as the ZnO concentration increases, the high surface energy of the nanoparticles promotes aggregation, which reduces their effective surface area and consequently decreases the availability of active hydrophilic sites. This diminished interaction between water molecules and the membrane surface counteracts the initial improvement in wettability, contributing to the decline in water flux observed at higher ZnO composition.

Fig. 3(b) illustrates the effect of ZnO composition on the permeability of PLA/ZnO membranes. Permeability peaks at 34.79 L/m²·h·bar at 0.25 wt.% ZnO, attributed to enhanced hydrophilicity and uniform nanoparticle distribution, optimizing water transport. However, at higher ZnO compositions (0.5–0.75 wt.%), permeability declines due to nanoparticle aggregation, leading to pore blockage and reduced effective pore size. Similarly, Shen et al. [23] reported similar trends, with permeability declining as ZnO concentrations rose due to irregular particle distribution and pore obstruction.

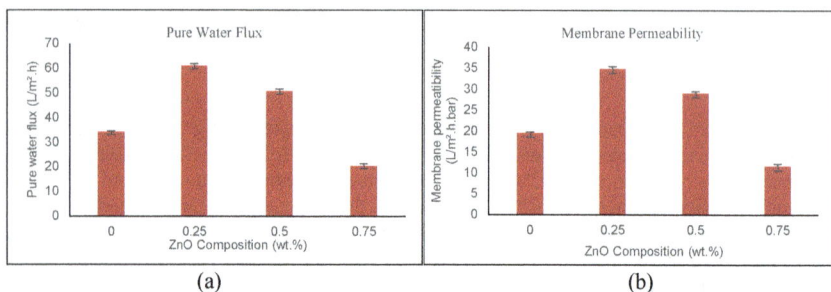

(a)   (b)

Fig. 3(a) Pure water flux & (b) Permeability of PLA/ZnO membranes

Dye Rejection:

Fig. 4(a) highlights the relationship between ZnO composition and the dye rejection rate of PLA membranes. Dye rejection improves with ZnO addition, peaking at 68.83% at 0.25 wt.% ZnO due to enhanced hydrophilicity and photocatalytic properties. However, increasing ZnO content beyond this optimal level leads to nanoparticle agglomeration, causing defects that reduce performance, with a rejection rate of 63.37% at 0.75 wt.% ZnO. This aligns with studies indicating that higher nanoparticle concentrations reduce the antifouling capabilities of membranes. Khan et al. [25] found MO dye removal efficiency of 55% in AC-PAA/PES membranes at 6 bar, but performance declined at higher pressures due to pore size reduction. Similarly, Mahmoudian and Kochameshki [26] observed a peak MO rejection of 92% in PES membranes with 1% GO-HBE, decreasing to 84% at 2% GO-HBE due to pore enlargement and reduced electrostatic repulsion.

Photocatalytic Reaction:

Fig. 4(b) illustrates the photocatalytic degradation efficiency of PLA/ZnO membranes with varying ZnO compositions. The pure PLA membrane (zero ZnO) shows a degradation efficiency of 64.79%, attributed to the absence of photocatalytic activity. At 0.25 wt.% ZnO, efficiency peaks at 71.90%, due to optimal nanoparticle dispersion, enhancing photocatalytic sites and light absorption. However, efficiency decreases to 67.80% at 0.5 wt.% ZnO and further to 66.49% at 0.75 wt.% ZnO, due to ZnO agglomeration, which reduces active sites and limits light penetration. Similarly, Hir et al. [11] found that PES/ZnO hybrid films achieved maximum degradation efficiency of 98% at 17 wt.% ZnO, with activity decreasing at higher concentrations due to nanoparticle agglomeration.

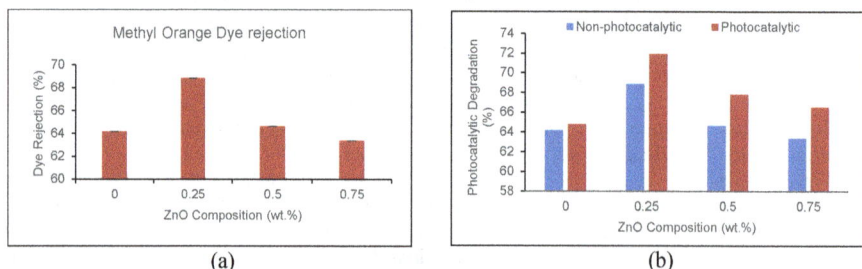

(a)   (b)

Fig. 4 (a) Dye rejection and (b) Photocatalytic reaction of PLA/ZnO membranes;

Frontiers of Chemical and Materials Engineering - ICoFCheM 2025 Materials Research Forum LLC
Materials Research Proceedings 60 (2026) 10-19 https://doi.org/10.21741/9781644903971-2

Conclusions

The incorporation of ZnO into PLA membranes significantly influenced their structural and functional properties. Increasing ZnO content improved hydrophilicity, as reflected by the reduction in water contact angle. Optimal porosity, flux, and permeability were obtained at moderate ZnO loading, beyond which agglomeration occurred, leading to pore blockage and reduced performance. Dye rejection and photocatalytic degradation were also enhanced at the optimal ZnO concentration, contributing to effective methyl orange removal. However, excessive ZnO addition diminished these benefits due to particle aggregation.

References

[1] R. Li, Y. Lou, Y. Xu, G. Ma, B. Liao, L. Shen, and H. Lin, Effects of surface morphology on alginate adhesion: Molecular insights into membrane fouling based on XDLVO and DFT analysis. Chemosphere, 233, 373–3 Alhoshan,0 (2019). https://doi.org/10.1016/j.chemosphere.2019.05.262

[2] H. M. Diallo, H. Ayyoub, F. Elazhar, M. Tahaikt, A. Elmidaoui, & M. Taky (2024). Evaluation of the potential ultrafiltration to improve the quality of secondary effluents in tertiary treatment and reuse. Physics and Chemistry of the Earth, Parts A/B/C, 135, 103672. https://doi.org/10.1016/j.pce.2024.103672

[3] H. Xu, K. Xiao, J. Yu, B. Huang, X. Wang, S. Liang, C. Wei, X. Wen, and X. Huang, A Simple Method to Identify the Dominant Fouling Mechanisms during Membrane Filtration Based on Piecewise Multiple Linear Regression. Membranes, 10(8), 171 (2020). https://doi.org/10.3390/membranes10080171

[4] N. H. Othman, N. H. Alias, N. S. Fuzil, F. Marpani, M. Z. Shahruddin, C. M. Chew, K. M. D. Ng, W. J. Lau, and A. F. Ismail, A review on the use of membrane technology systems in developing countries. Membranes, 12(1), 30 (2021). https://doi.org/10.3390/membranes12010030

[5] E. H. Alosaimi, H. M. Hassan, I. H. Alsohaimi, Q. Chen, S. Melhi, A. A. Younes, and W. H. El-Shwiniy, Fabrication of sulfonated polyethersulfone ultrafiltration membranes with an excellent antifouling performance by impregnating with polysulfopropyl acrylate coated ZnO nanoparticles. Environmental Technology & Innovation, 25, 102210 (2021). https://doi.org/10.1016/j.eti.2021.102210

[6] Mezher, H. M., Adeli, H., & Alsalhy, Q. F. (2024). Novel ZnO-modified Polyethersulfone nanocomposite membranes for Nanofiltration of concentrated textile wastewater. Water, Air, & Soil Pollution, 235(2), 138.https://doi.org/10.1007/s11270-024-06927-7

[7] M. Alhoshan, J. Alam, L. A. Dass, and N. Al-Homaidi, Fabrication of Polysulfone/ZNO membrane: Influence of ZNO nanoparticles on membrane characteristics. Advances in Polymer Technology, 32(4) (2013). https://doi.org/10.1002/adv.21369

[8] A. Iulianelli, F. Russo, F. Galiano, M. Manisco, and A. Figoli, Novel bio-polymer based membranes for CO2/CH4 separation. International Journal of Greenhouse Gas Control, 117, 103657 (2022). https://doi.org/10.1016/j.ijggc.2022.103657

[9] K. M. Lee, C. W. Lai, K. S. Ngai, and J. C. Juan, Recent developments of zinc oxide based photocatalyst in water treatment technology: A review. Water Research, 88, 428–448 (2016). https://doi.org/10.1016/j.watres.2015.09.045

[10] Huang, S. Si, Z. Chen, & B. Xin, (2025). Preparation, characterization and antibacterial evaluation of PLA/ZnO nanofiber membranes loaded with thymol by coaxial electrospinning.

Journal of Drug Delivery Science and Technology, 107040.
https://doi.org/10.1016/j.jddst.2025.107040

[11] Z.M. Hir, A. Abdullah, Z. Zainal, and H. Lim, Photoactive hybrid film photocatalyst of polyethersulfone-ZnO for the degradation of methyl orange dye: Kinetic study and operational parameters. Catalysts, 7(11), 313 (2017). https://doi.org/10.3390/catal7110313

[12] S. Zhao, W. Yan, M. Shi, Z. Wang, J. Wang, and S. Wang, Improving permeability and antifouling performance of polyethersulfone ultrafiltration membrane by incorporation of ZnO-DMF dispersion containing nano-ZnO and polyvinylpyrrolidone. Journal of Membrane Science, 478, 105–116 (2015). https://doi.org/10.1016/j.memsci.2014.12.050

[13] M. F. Ismail, M. A. Islam, B. Khorshidi, A. Tehrani-Bagha, and M. Sadrzadeh, Surface characterization of thin-film composite membranes using contact angle technique: Review of quantification strategies and applications. Advances in Colloid and Interface Science, 299, 102524 (2022). https://doi.org/10.1016/j.cis.2021.102524

[14] T. Xiao, P. Wang, X. Yang, X. Cai, and J. Lu, Fabrication and characterization of novel asymmetric polyvinylidene fluoride (PVDF) membranes by the nonsolvent thermally induced phase separation (NTIPS) method for membrane distillation applications. Journal of Membrane Science, 489, 160–174 (2015). https://doi.org/10.1016/j.memsci.2015.03.081

[15] Nasib, A. M.; Baharudin, N.; Jullok, N.; Rasidi, S.; and Jaafar, J. (2023). Fabrication of biodegradable Polylactic Acid (PLA) membrane for reverse osmosis process. AIP Conference Proceedings. https://doi.org/10.1063/5.0115854

[16] Y. K. Poon, S. K. E. A. Rahim, Q. H. Ng, P. Y. Hoo, N. Y. Abdullah, A. Nasib, and N. S. Abdullah, Synthesis and characterisation of Self-Cleaning TIO2/PES mixed matrix membranes in the removal of humic acid. Membranes, 13(4), 373 (2023). https://doi.org/10.3390/membranes13040373

[17] T. Da Silva Neto, L. S. Maia, M. O. T. Da Conceição, M. B. Da Silva, L. T. Carvalho, S. F. Medeiros, M. I. S. D. Faria, B. B. Migliorini, R. Lima, D. S. Rosa, and D. R. Mulinari, Enhancing PLA filament biocompatibility by introducing ZnO and ketoprofen. Journal of Inorganic and Organometallic Polymers and Materials (2024). https://doi.org/10.1007/s10904-024-03275-1

[18] J. Ding, Z. Mao, H. Chen, X. Zhang, and H. Fu, Fabrication of ZnO/PDA/GO composite membrane for high efficiency oil–water separation. Applied Physics A, 129(5) (2023). https://doi.org/10.1007/s00339-023-06654-6

[19] D. A. Goncharova, E. N. Bolbasov, A. L. Nemoykina, A. A. Aljulaih, T. S. Tverdokhlebova, S. A. Kulinich, and V. A. Svetlichnyi. Structure and Properties of Biodegradable PLLA/ZnO Composite Membrane Produced via Electrospinning. Materials, 14(1), 2 (2020). https://doi.org/10.3390/ma14010002

[20] Z. Tang, F. Fan, Z. Chu, C. Fan, and Y. Qin, Barrier Properties and Characterizations of Poly(lactic Acid)/ZnO Nanocomposites. Molecules, 25(6), 1310 (2020). https://doi.org/10.3390/molecules25061310

[21] V. Salaris, I. S. F. García-Obregón, D. López, and L. Peponi, Fabrication of PLA-Based Electrospun Nanofibers Reinforced with ZnO Nanoparticles and In Vitro Degradation Study. Nanomaterials, 13(15), 2236 (2023). https://doi.org/10.3390/nano13152236

[22] M. Murali, H. G. Gowtham, N. Shilpa, S. B. Singh, M. Aiyaz, R. Z. Sayyed, C. Shivamallu, R. R. Achar, E. Silina, V. Stupin, N. Manturova, A. A. Shati, M. Y. Alfaifi, S. E. I. Elbehairi,

and S. P. Kollur, Zinc Oxide Nanoparticles Prepared through Microbial Mediated Synthesis for Therapeutic Applications: A Possible Alternative for Plants. Frontiers in Microbiology, 14 (2023). https://doi.org/10.3389/fmicb.2023.1227951

[23] L. Shen, X. Bian, X. Lu, L. Shi, Z. Liu, L. Chen, Z. Hou, and K. Fan, Preparation and Characterization of ZnO/Polyethersulfone (PES) Hybrid Membranes. Desalination, 293, 21–29 (2012). https://doi.org/10.1016/j.desal.2012.02.019

[24] P. Moradihamedani, N. A. Ibrahim, D. Ramimoghadam, W. M. Z. W. Yunus, and N. A. Yusof, Polysulfone/Zinc Oxide Nanoparticle Mixed Matrix Membranes for CO2/CH4 Separation. Journal of Applied Polymer Science, 131(16) (2013). https://doi.org/10.1002/app.39745

[25] I. A. Khan, K. M. Deen, E. Asselin, M. Yasir, R. Sadiq, and N. M. Ahmad, Boosting Water Flux and Dye Removal: Advanced Composite Membranes Incorporating Functionalized AC-PAA for Wastewater Treatment. Journal of Industrial and Engineering Chemistry, 145, 705-720 (2024). https://doi.org/10.1016/j.jiec.2024.10.067

[26] M. Mahmoudian and M.G. Kochameshki, The performance of polyethersulfone nanocomposite membrane in the removal of industrial dyes. Polymer, 224, 123693 (2021). https://doi.org/10.1016/j.polymer.2021.123693

Frontiers of Chemical and Materials Engineering - ICoFCheM 2025   Materials Research Forum LLC
Materials Research Proceedings 60 (2026) 20-25   https://doi.org/10.21741/9781644903971-3

Optimization of GMAW Parameters for Marine-Grade Aluminium Alloy 5083-H116: Weld Quality and Mechanical Performance Assessment

M.F.M. WAHID[1,2,a*], N.H.F. HAMZAH[1,b], S.R. SHAMSUDIN[1,c], M. MUSA[1,d]

[1]Faculty of Mechanical Engineering & Technology, Universiti Malaysia Perlis, 02600 Arau, Perlis
Malaysia

[2]Frontier Materials Research, Centre of Excellence (FrontMate), Universiti Malaysia Perlis
(UniMAP), 02600 Arau, Perlis, Malaysia

[a]fitri@unimap.edu.my, [b]fahmihamzah9998@gmail.com, [c]rizam@unimap.edu.my,
[d]mmlan456456@gmail.com

Keywords: GMAW, Aluminium 5083-H116, Welding Parameter Optimization, Marine Structures, Non-Destructive Testing, Tensile Properties

Abstract. The demand for lightweight and corrosion-resistant materials in shipbuilding has led to the widespread use of aluminium alloys, particularly 5083-H116. Gas Metal Arc Welding (GMAW) is commonly used for joining these alloys due to its versatility and productivity. However, selecting optimal process parameters is critical to achieving defect-free welds with desirable mechanical properties. This study examines the impact of varying welding current and voltage on weld quality in aluminum alloy 5083-H116. Welds were assessed using dye penetrant inspection and tensile tests in both longitudinal and transverse orientations. The results showed that high current (170 A) and corresponding voltage (22 V) offered superior performance, reducing surface defects and improving mechanical strength. These findings serve as a practical guideline for improving GMAW application in aluminium shipbuilding, promoting structural reliability, and minimizing rework or failure risk.

Introduction

The marine industry faces increasing demand for efficient, durable, and corrosion-resistant vessels [1-3]. In response, shipbuilders frequently turn to aluminium alloy 5083-H116, a high-strength, marine-grade alloy known for its corrosion resistance and weldability [4]. However, welding aluminium presents unique challenges. Gas metal arc welding (GMAW) of aluminium and marine-grade aluminium alloys is highly sensitive to variations in current, voltage, welding speed, and overall heat input, all of which critically determine weld quality and defect formation. Excessive current or heat input can result in hot cracking, burn-through, and undercutting along the weld toe due to excessive melting and steep thermal gradients, whereas insufficient current leads to lack of fusion, cold laps, or weak penetration. Improper voltage control causes arc instability, which promotes spatter, porosity. Likewise, high welding speeds lower heat input and induce incomplete penetration and shrinkage cavities, while slow travel speeds increase heat accumulation, resulting in distortion, grain coarsening, and a wider heat-affected zone (HAZ). These combined effects can also lead to micro-voids and oxide inclusions, especially when gas shielding or surface preparation is inadequate. Therefore, maintaining optimal control over current, voltage, speed, and heat input is essential in GMAW to ensure stable arc conditions, uniform bead geometry, and the production of defect-free welds with reliable mechanical performance in both general and marine aluminium structures [5,6].

Welded aluminium alloys using GMAW generally exhibit lower tensile strength compared to their base metals. This reduction in strength is primarily due to the formation of porosity, grain coarsening, and the loss of strengthening second-phase particles during welding. AA5059 alloy

Frontiers of Chemical and Materials Engineering - ICoFCheM 2025   Materials Research Forum LLC
Materials Research Proceedings 60 (2026) 20-25   https://doi.org/10.21741/9781644903971-3

welded using GMAW shows a 30% reduction in tensile strength compared to GTAW joints, which perform better due to finer grain structures and lower heat input [7,8].

While previous studies have provided general guidelines for welding aluminium [9,10], relatively few focus specifically on optimizing GMAW parameters for 5083-H116 in a shipbuilding context. This study aims to bridge that gap by providing experimental insights on how different parameter combinations affect weld integrity and mechanical performance.

Materials & Methodology
The base material used in this study was aluminium alloy 5083-H116, supplied in 6 mm thick plates. This alloy is known for its strength in marine environments and is commonly used in hull construction and superstructures. The filler metal was ER5183 aluminium wire, selected for its compatibility with 5xxx-series aluminium alloys and excellent corrosion resistance. Table 1 shows the chemical composition of the base metal (aluminium alloy 5083-H6115) and filler (ER5183), respectively.

Table 1: Chemical composition of base metal, aluminium alloy 5083-H116 and filler material, ER5138

	Mg	Mn	Si	Fe	Zn	Ti	Cu	Cr	Al
5083-H116	4.0 – 4.9	0.4 – 1.0	Max. 0.4	Max. 0.4	Max. 0.25	Max. 0.25	Max. 0.10	0.05 – 0.25	Bal.
ER5138	4.3 – 5.2	0.5 – 1.0	Max. 0.4	Max. 0.4	Max. 0.25	Max. 0.15	Max. 0.10	0.05 – 0.25	Bal.

Welding was conducted using the MEGMEET Ehave C500 digital inverter GMAW power source. Pure argon was used as the shielding gas at a flow rate of 20 L/min to prevent oxidation and contamination. Travel speed and wire feed rate were fixed at 11.3mm/sec and 8.2mm/min, respectively. Welding was performed in the flat position using a double-V groove configuration. Three welding parameter sets were developed to investigate the effects of current: (a) low current, (b) medium current, and (c) high current. Table 2 shows the three parameter sets for this study.

Table 2: Welding parameters for alloy 5083-H116 using MEGMEET Ehave C500 digital inverter GMAW

Parameter set	Filler size [mm]	Current [A]	Voltage [V]	Wire speed [mm/min]	Travel speed [mm/sec]	Gas flow [l/min]	Polarity
Low current	1.2	150	21	8.2	11.3	20	DCEP
Medium current	1.2	160	22	8.2	11.3	20	DCEP
High current	1.2	170	22	8.2	11.3	20	DCEP

Each welded plate was sectioned into test coupons. Surface cleaning was performed using a wire brush and acetone to remove oxides and oil residues. The cleaned specimens were subjected to both non-destructive and mechanical tests to evaluate weld quality and strength. Visual inspection was performed after welding was carried out. Dye penetrant test (DPT) as accordance to AWS 1.10 Guide for Nondestructive Examination of Welds was used to detect surface-level weld defects such as cracks, porosity, or incomplete fusion. A red dye was applied, followed by a developer spray to highlight defect locations. In the destructive test, rectangular-shaped tensile

Frontiers of Chemical and Materials Engineering - ICoFCheM 2025   Materials Research Forum LLC
Materials Research Proceedings 60 (2026) 20-25   https://doi.org/10.21741/9781644903971-3

specimens were prepared according to AWS 4.0 Standard Methods for Mechanical Testing for Welds. Tests were conducted using a universal testing machine according to ASTM E8 Standard Test Methods for Tension Testing for Metallic Materials. Ultimate tensile strength (UTS), yield strength (YS), and elongation were recorded for both longitudinal and transverse weld orientations. Figure 1 shows the transverse tensile specimen in accordance to AWS 4.0 standard.

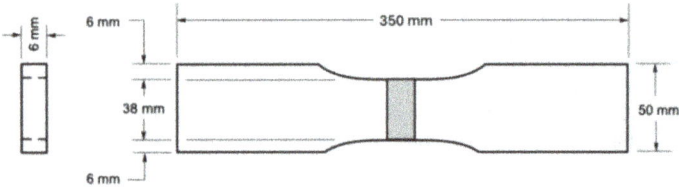

Figure 1: Transverse tensile specimen

Results & Discussion

Surface analysis revealed a clear correlation between welding parameters and defect occurrence. For welds produced under low current conditions (150 A), the somewhat linear and intermittent nature of the red indication along the weld bead as shown in Fig. 2(a) suggests a lack of fusion between the weld metal and the base metal, or between weld passes. Weld generated under medium current conditions (160 A), as shown in Fig. 2(b), on the other hand, showed signs of spatter and porosity. A diffuse, almost smeared red indication running along the length of the weld bead was observed for high current condition (170 A) as shown in Fig. 2(c). The widespread, somewhat hazy red-bleed-out could indicate the presence of very fine, distributed porosity bear the surface. The dye test in Figures 1 confirmed the absence of major surface-breaking defects, suggesting optimal fusion and arc control at this range [11].

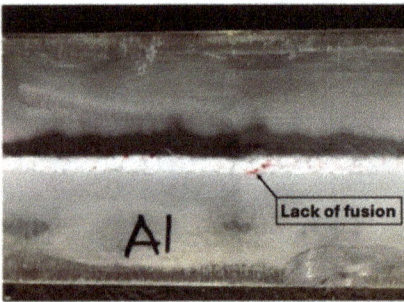
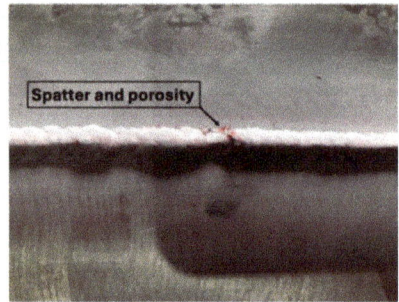

(a)   (b)

Frontiers of Chemical and Materials Engineering - ICoFCheM 2025
Materials Research Proceedings 60 (2026) 20-25

Materials Research Forum LLC
https://doi.org/10.21741/9781644903971-3

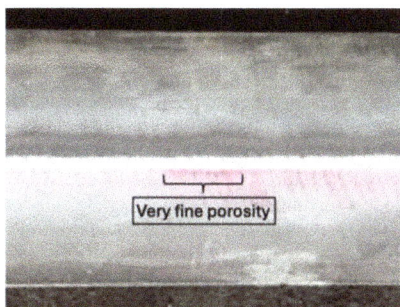

(c)

Figure 2: Dye penetrant test result for (a) low, (b) medium, and (c) high current conditions

Sample that was welded at 170 A shows the highest longitudinal strength of 100.521 MPa with moderate elongation of 17%. These highest tensile strength and moderate elongation were achieved at high current indicating effective penetration and minimized heat-affected zone (HAZ) degradation [12]. The transverse strength of 63.506 MPa, although lower than longitudinal due to weld orientation, also followed the same trend.

Table 3: Mechanical testing results

Parameter Set	UTS (Longitudinal) [MPa]	UTS (Transverse) [MPa]	Elongation (Longitudinal) [%]
Low Current	96.354	61.234	12.5
Med Current	77.083	60.713	21.3
High Current	100.521	63.506	17.0

The surface analysis conducted on the 5083-H116 aluminium alloy welds produced using GMAW revealed a distinct correlation between the applied welding parameters and the nature of defect occurrence. Under low current conditions, the dye penetrant test highlighted linear and intermittent red indications along the weld bead, strongly suggesting a lack of adequate fusion between the weld metal and the base material, or between subsequent weld passes. This observation aligns with the understanding that insufficient heat input at lower current levels hinders proper melting and the establishment of a robust metallurgical bond [13,14].

Conversely, welds generated under medium current conditions exhibited signs of both spatter and porosity. This suggests that the parameters employed at this medium current range likely induced increased turbulence within the molten weld pool, facilitating the entrapment of gases and the subsequent formation of pores.

Interestingly, the application of high current during welding resulted in a diffuse, almost smeared red indication extending along the length of the weld bead. This widespread, hazy bleed-out is indicative of very fine, distributed porosity located near the surface. Despite this observation, the dye penetrant test confirmed the absence of any major surface-breaking defects, implying that optimal fusion and arc control were achieved within this high current. This finding was further corroborated by the tensile testing results, which demonstrated that the highest tensile strength of 100.521 MPa for longitudinal welded joint and moderate elongation of 17% were achieved under high current conditions (170 A). This superior mechanical performance suggests effective penetration and a minimized degradation of the heat-affected zone. The transverse tensile strength of 63.506 MPa, while expectedly lower than the longitudinal strength due to the inherent

Frontiers of Chemical and Materials Engineering - ICoFCheM 2025
Materials Research Proceedings 60 (2026) 20-25

Materials Research Forum LLC
https://doi.org/10.21741/9781644903971-3

anisotropy of the welded structure, followed the same trend, indicating a consistent influence of welding parameters on the overall mechanical integrity of the weld. In contrast, low current conditions (150 A) yielded the lowest tensile strength of 96.354 MPa and elongation of 12.5%, while medium current conditions resulted in intermediate mechanical properties.

These findings collectively emphasize the critical role of welding parameters in determining the quality and performance of 5083-H116 aluminium alloy welds produced by GMAW. The lack of fusion observed at low current directly translates to compromised mechanical strength. Similarly, the porosity evident at medium current, while perhaps not severe surface-breaking, still decreases from the overall integrity. The high current condition, despite the presence of fine, distributed porosity near the surface, ultimately delivered the best balance of fusion and mechanical properties. This suggests that while high current facilitates optimal penetration and minimizes HAZ issues, further investigation into reducing the formation of this fine porosity could potentially lead to even greater improvements in weld quality. The distinct trends observed in both non-destructive testing and mechanical evaluation provide a solid foundation for further discussion regarding the optimization of welding parameters for this specific alloy and welding process.

Conclusion

This study demonstrates that among the tested welding parameters, a high current setting of 170 A yields the most favourable outcomes for GMAW of 5083-H116 aluminium alloy. The results confirm that this parameter facilitates optimal fusion and arc stability, effectively minimizing the occurrence of major surface-breaking defects as verified through dye penetrant testing. Furthermore, the mechanical performance under this condition—characterized by enhanced tensile strength and acceptable elongation—indicates improved metallurgical bonding and reduced heat-affected zone degradation. While fine, distributed porosity was observed near the surface, it did not significantly compromise weld integrity. These findings suggest that high current welding is the most suitable for achieving robust and defect-free joints in this alloy system. Future work may focus on refining this parameter to further mitigate minor porosity and enhance overall weld quality.

Acknowledgement

The author gratefully acknowledges the support of Gading Marine Industry for providing materials and testing facilities.

References

[1] W. Gao, D. Lu, H. Wang, Q. Jiang, J. Li, C. Li, Y. Huang, The development and utilization of ecofriendly and high-performance materials in the marine environment, Adv. Eng. Mater., 27 (2025). https://doi.org/10.1002/adem.202500149

[2] M. Rezayat, M. Karamimoghadam, M.L. Dezaki, A. Zolfagharian, G. Casalino, A. Mateo, M. Bodaghi, Enhancing corrosion resistance and mechanical strength of 3D-printed iron polylactic acid for marine applications via laser surface texturing, Adv. Eng. Mater., 2402591 (2024). https://doi.org/10.1002/adem.202402591

[3] G.P.V. Dalmora, E.P.B. Filho, A.A.M. Conterato, W.S. Roso, C.E. Pereira, A. Dettmer, Methods of corrosion prevention for steel in marine environments: A review, Results Surf. Interf., 18 (2025) 100430. https://doi.org/10.1016/j.rsurfi.2025.100430

[4] M. Umar, G. Balaji, A. Maria Jackson, S. Jayasathyakawin, Microstructure and corrosion characteristics of AA5083 alloy weld beads produced by GTAW and SpinArc-GMAW, Mater. Today Commun., 41 (2024) 110467. https://doi.org/10.1016/j.mtcomm.2024.110467

[5] M. Dada, P. Popoola, Recent advances in joining technologies of aluminum alloys: A review, Discov. Mater., 4 (2024) 86. https://doi.org/10.1007/s43939-024-00155-w

[6] B.-Q. Chen, K. Liu, S. Xu, Recent advances in aluminum welding for marine structures, J. Marine Sci. Eng., 12 (2024) 1539. https://doi.org/10.3390/jmse12091539

[7] L. Huang, X. Hua, D. Wu, Z. Jiang, Y. Ye, A study on the metallurgical and mechanical properties of a GMAW-welded Al-Mg alloy with different plate thicknesses, J. Manufac. Proc., 37 (2019) 438–445

[8] K. Vasu, H. Chelladurai, A. Ramaswamy, S. Malarvizhi, V. Balasubramanian, Effect of fusion welding processes on tensile properties of armor grade, high thickness, non-heat treatable aluminium alloy joints, Def. Tech., 15 (2019) 353e362

[9] L.-j. Wu, X.-h. Han, G.-l. Ma, B. Yang, H. Bian, X.-g. Song, C.-w. Tan, Effects of welding layer arrangement on microstructure and mechanical properties of gas metal arc welded 5083/6005A aluminium alloy butt joints, Trans. Nonferrous Met. Soc. China, 3 (2023) 1665-1676. https://doi.org/10.1016/S1003-6326(23)66212-0

[10] B. M. Sharma, T. Bajpai, P. K. Gupta, Vikash Gautam, Effect of operating parameters of hybrid TIG-MIG welding on mechanical properties and weld bead quality: A review, Mater. Today Proc., (2023). https://doi.org/10.1016/j.matpr.2023.01.325

[11] M. Dekis, M. Tawfik, M. Egiza, M. Dewidar, Challenges and developments in wire arc additive manufacturing of steel: A review, Res. Eng., 26 (2025) 104657

[12] S. Yang, X. Yang, X. Lu, M.V. Li, H. Zuo, Y. Wang, Strength calculation and microstructure characterization of HAZ softening area in 6082-T6 aluminum alloy CMT welded joints, Mater. Today Comm., 37 (2023), 107077. https://doi.org/10.1016/j.mtcomm.2023.107077

[13] S. Selvamani, M. Bakkiyaraj, V. Balasubramanian, A. Ramesh, M. Vijayakumar, Investigation of heat input effects on the joint characteristics of CMT welded AA6061 sheets, Proc. Inst. Mech. Eng. C: J. Mech. Eng. Sci., 237 (2022) 2328-2345. https://doi.org/10.1177/09544062221138839

[14] W. Guo, X. Zhao, Y. Zhao, Y. Liu, H. Li, B. Han, Research on optimal heat input parameter for TIG welding of thin plate 5083 aluminum alloy, Sci. Rep., 15(2025) 15593. https://doi.org/10.1038/s41598-025-99836-6

Frontiers of Chemical and Materials Engineering - ICoFCheM 2025
Materials Research Proceedings 60 (2026) 26-33

Materials Research Forum LLC
https://doi.org/10.21741/9781644903971-4

Impact of Alkali Treatment on the Morphology, Mechanical, and Moisture Characteristics of Banana Fiber

Siti Aisyah binti Azman[1,a] *, Ahmad Humaizi bin Hilmi[1,b],
Asna Rasyidah binti Abdul Hamid[1,c] and Abdul Rashid bin Othman[1,d]

[1]Faculty of Mechanical Engineering & Technology, Universiti Malaysia Perlis, Arau, Perlis, Malaysia

[a]aisyahazman@studentmail.unimap.edu.my, [b]humaizi@unimap.edu.my, [c]asnarasyidah@unimap.edu.my, [d]rashidothman@studentmail.unimap.edu.my

Keywords: Alkali Treatment, Banana Fiber, Fiber Crystallinity, Mechanical Properties, Moisture Absorption

Abstract. This study investigates the influence of alkali treatment on the morphological, mechanical, and moisture properties of banana fibers, emphasizing their potential for sustainable composite applications. Treating banana fibers with sodium hydroxide (NaOH) removed non-cellulosic constituents such as lignin, hemicellulose, and waxes, thereby modifying surface roughness, tensile performance, and hydrophilicity. Concentrations of 1%, 3%, 5%, and 7% NaOH were evaluated using tensile testing, Optical Microscope (OM), Scanning Electron Microscopy (SEM) and moisture measurement. Among these, 5% NaOH treatment produced the most optimum outcomes, with tensile strength increasing by approximately 52% and elongation at break by nearly 69% compared to untreated fibers. Moisture absorption decreased by about 39%, enhancing dimensional stability. Higher concentrations caused fiber degradation and reduced mechanical performance. These findings highlight 5% NaOH as the optimal treatment for improving fiber–matrix compatibility while maintaining structural integrity, supporting the development of banana fiber composites for engineering applications.

Introduction

Banana fibers, derived from the pseudostem of Musa species, are being investigated as reinforcement in composites owing to their availability, high cellulose content, biodegradability, and classification as an agricultural byproduct [1]. The use of these fibers minimizes waste and supports sustainable material development. In comparison to other natural fibers like jute, sisal, and hemp, banana fibers exhibit a higher cellulose content and reduced processing costs, rendering them appropriate for extensive eco-friendly applications [2,3]. Untreated fibers exhibit significant moisture absorption, inadequate interfacial bonding, and irregular morphology, which restrict their structural performance [4].

Alkali treatment using sodium hydroxide (NaOH), referred to as mercerization, is commonly utilized to enhance fiber properties through the removal of hemicellulose and lignin, which exposes cellulose microfibrils and increases crystallinity [5]. Chemical modifications enhance surface roughness, facilitate fiber–matrix adhesion, and reduce hydrophilicity [6]. Although optical microscopy (OM) is proficient for fundamental morphological examination, it is unable to resolve nanoscale characteristics. SEM offers high-resolution observations of fibrillation and surface cleaning [7].

Recent studies indicate that alkali treatment improves the mechanical strength, flexibility, and moisture resistance of natural fiber composites [8–11]. The findings confirm that treatment effectiveness is contingent upon concentration, with moderate NaOH levels yielding optimal improvements [12,13]. Advanced investigations employing optimization techniques confirmed that controlled alkali treatment enhances strength while maintaining fiber integrity [14,15].

Frontiers of Chemical and Materials Engineering - ICoFCheM 2025   Materials Research Forum LLC
Materials Research Proceedings 60 (2026) 26-33   https://doi.org/10.21741/9781644903971-4

However, few studies have systematically investigated banana fibers at various NaOH concentrations while concurrently assessing morphology, tensile properties, and moisture content. This research fills the existing gap by utilizing OM, SEM, mechanical testing, and moisture analysis to determine the optimal NaOH concentration for the treatment of banana fiber.

Material and Method

Material. The primary material used in this study, banana pseudo-stem, was sourced from farms in Kedah, Malaysia. This agricultural byproduct was selected for its natural abundance and high cellulose content, making it suitable for eco-friendly applications. Sodium hydroxide (NaOH), required for the alkali treatment, was procured from Scientific Supply S. The pseudo-stem was carefully prepared by removing impurities to ensure consistent quality for testing. The NaOH solution was used to modify the fiber's surface, improving its mechanical and chemical properties by eliminating lignin, hemicellulose, and other unwanted components.

Banana Fiber Preparation. The pseudo-stem of the banana plant was processed by removing outer sheaths and discarding damaged layers. Inner sections were cut into uniform 50 mm × 10 mm pieces, washed repeatedly to remove impurities, and air-dried at room temperature for 24 hours to prepare for further treatment and analysis.

Alkali Treatment of Banana Fiber. Banana fibers were treated with NaOH solutions (1%, 3%, 5%, 7%) at a 1:20 fiber-to-liquid ratio for 1 hour at room temperature (Figure 1). Fibers were rinsed with distilled water and air-dried for 48 hours. This process improves surface properties. Chemical interaction is shown in Equation (1).

Figure 1: Banana fibers were soaked in NaOH solution.

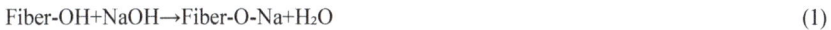

$$Fiber\text{-}OH + NaOH \rightarrow Fiber\text{-}O\text{-}Na + H_2O \tag{1}$$

Morphology Analysis. Morphology was observed by Optical Microscope (OM) at 5× and 15× magnification and by Scanning Electron Microscopy SEM (Hitachi TM3000 Tabletop Scanning Electron Microscope) at 50 –100× magnification. OM was used for basic surface evaluation, while SEM provided detailed characterization of fibrillation, impurity removal, and microstructural changes [7,12].

Mechanical Testing. Tensile strength and elongation at break were evaluated according to ASTM D638-4 utilizing a 50 kN Shimadzu Universal Testing Machine, operated with Trapezium X software at a speed of 10 mm/min. Dumbbell-shaped samples measuring 115 mm × 19 mm × 4 mm were prepared from both untreated and treated banana fibers, as illustrated in Figure 2.

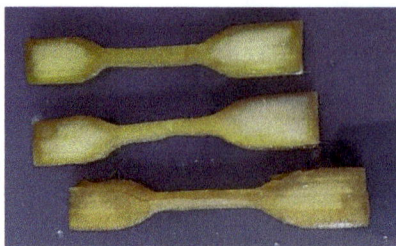

Figure 2: Banana fibers were cut into dumbbell shapes.

The tensile strength and elongation at break, respectively, were calculated by the following Equation 2 and Equation 3:

Tensile strength, Mpa,

$$s = P/a \qquad (2)$$

where:
s = stress
P = force required to break
a = cross-sectional area

Elongation at break, %,

$$\varepsilon = (\Delta L/L) \times 100 \qquad (3)$$

where:
ε = elongation
ΔL = Final length
L = Initial length

Moisture Test. Moisture contents of untreated and alkali-treated fibers were obtained using electrical resistance and capacitance-based non-destructive techniques with handheld moisture meter [5,6]. Fibers were treated with 1%, 3%, 5%, and 7% NaOH solutions as well as control (untreated) and rinsed free of residual chemicals. The samples were air-dried at room temperature for 42 hours. A tiny bit of each dried sample was examined by sending an electrical current through the fibers. The device detected differences in conductivity or capacitance as a function of moisture content. Each sample was read 3 times, and average values were determined to ensure reliable data for this study.

Results and Discussion
Optical Microscope (OM). There were noticeable structural variations between the untreated and alkali-treated banana fibers, according to the morphological study. Figure 3(a) shows that untreated fibers had pores that were visible, which is a reflection of their porous structure. In contrast, Figure 3(b) shows that fibers treated with 5% NaOH had smoother, non-porous surfaces because hemicellulose, lignin, pectin, and waxes were removed [12]. In addition to increasing homogeneity, alkali treatment generated fibrillation, which divided bundles into finer fibers [8]. Surface improvement was negligible at 1% and 3% concentrations, and fiber damage occurred at

7% concentration and above. Composite compatibility and overall structural performance were both improved by the 5% NaOH treatment, which produced the best possible surface quality [9, 13].

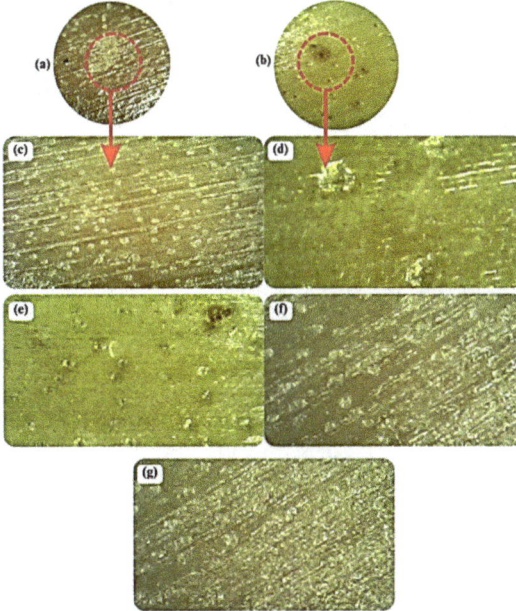

Figure 3: 5x magnification; (a) untreated banana fiber; (b) 5% NaOH-treated banana fiber; 15x magnification for (c) untreated, (d) 5%, (e) 1%. (f) 3%, (g) 7% NaOH banana fiber.

Scanning Electron Microscopy (SEM). Figure 4 presents SEM images of banana fibers subjected to different NaOH concentrations (1%, 3%, 5%, 7%, and 9%). At 1% NaOH, the fiber surface exhibited a significant presence of impurities and waxy layers, suggesting an incomplete removal of non-cellulosic materials. At 3% NaOH, partial fibrillation and cleaner surfaces were noted; however, some residues remained, indicating only moderate removal of hemicellulose.

The fibers treated with 5% NaOH demonstrated the most notable enhancement, characterized by distinctly fibrillated surfaces, decreased voids, and the exposure of finer cellulose fibrils. This demonstrates the successful elimination of hemicellulose and lignin, enhancing surface roughness and the potential for fiber–matrix adhesion. At 7% NaOH, over-etching effects were evident, resulting in surface damage and the emergence of cracks, which compromised structural integrity. At 9% NaOH, significant degradation was observed, characterized by fiber fragmentation, fibril pull-outs, and a loss of cohesive structure, indicating that high alkali concentration compromises fiber quality.

SEM observations indicate that a 5% NaOH solution achieves an optimal balance between surface cleaning and the preservation of structural integrity. Concentrations exceeding this threshold lead to over-etching, thereby elucidating the decrease in tensile strength and the heightened moisture retention noted in mechanical and moisture assessments. The findings align with earlier studies indicating that mid-range alkali treatments yield optimal reinforcement performance in natural fiber composites [9,12,14].

Frontiers of Chemical and Materials Engineering - ICoFCheM 2025   Materials Research Forum LLC
Materials Research Proceedings 60 (2026) 26-33   https://doi.org/10.21741/9781644903971-4

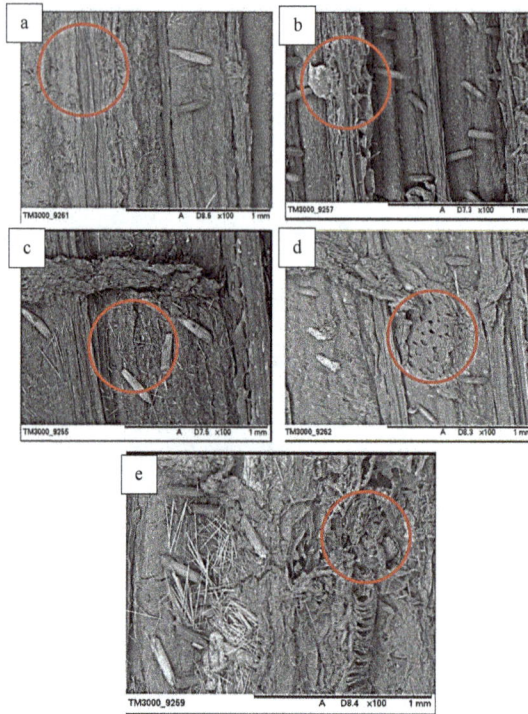

Figure 4: SEM micrographs of banana fibers treated with NaOH: (a) 1%, (b) 3%, (c) 5%, (d) 7%, (e) 9%.

Mechanical Testing. By effectively removing lignin, 5% NaOH improved cellulose alignment and load distribution, leading to a notable increase in tensile strength and elongation at break. Enhanced performance was negligible at concentrations between 1% and 3%, and fiber deterioration due to over-treatment occurred at 7%, lowering mechanical characteristics and structural integrity.

Table 1 shows the mechanical qualities of banana fiber are improved by carefully managed NaOH treatment. At 5% NaOH, tensile strength increases from 2.9561 MPa (untreated) to 4.5168 MPa because lignin and hemicellulose are optimally removed without causing fiber damage. Overtreatment results in a loss in strength at 7% NaOH (3.9854 MPa). Better flexibility is indicated by an increase in elongation at break from 31.88% (untreated) to 53.98% at 5% NaOH. At 7% NaOH, a drop to 49.93% indicates fiber degradation. Superior tensile and elongation performance results from the optimal balance between stiffness removal and fiber integrity provided by the 5% NaOH treatment. Excess alkali reduces the efficacy of fiber by disrupting its structure. Higher elongation indicates improved flexibility and toughness, enhancing energy absorption and making fibers more suitable for impact-resistant composites [11,13].

The increase in elongation at break is notably significant for practical composite applications. Fibers subjected to 5% NaOH treatment demonstrated elongation values of approximately 54%, in contrast to roughly 32% for untreated fibers. This enhancement suggests that the fibers can sustain increased deformation prior to fracture, indicating improved ductility and energy absorption capacity. In a composite system, increased elongation at break correlates with enhanced toughness, enabling the material to absorb greater energy during impact or cyclic loading. This property is essential in applications including automotive interior panels, protective gear, and packaging materials, where resistance to cracking and sudden failure is necessary. The combination of increased tensile strength and elongation observed at 5% NaOH enhances both stiffness and toughness, rendering treated banana fibers more appropriate for structural and impact-resistant applications.

Table 1: Tensile properties of untreated and treated banana fiber with different concentrations.

Sample	Stress, MPa	Elongation at break, %
Untreated	2.9561	31.8798
1% NaOH	3.0703	35.7232
3% NaOH	3.0885	39.9671
5% NaOH	4.5168	53.9783
7% NaOH	3.1854	49.9260

Moisture Test. Table 2 illustrates how the moisture content of banana fibers drops as the concentration of NaOH rises. Because of the high levels of hemicellulose and lignin, untreated fibers retain 49.93% of their moisture content. The moisture content decreases to 44.34% at 1% NaOH, indicating a partial elimination of hydrophilic components. Optimized elimination of moisture-retaining compounds is indicated by a further decline to 40.43% at 3% and a notable reduction to 30.53% at 5%. But at 7%, the moisture content rises to 35.80%, most likely as a result of lignin and hemicellulose being removed excessively, exposing more cellulose. This implies that the best solution for reducing moisture content without sacrificing fiber structure is 5% NaOH.

Table 2: The moisture test result after 42 hours

Sample	Test 1	Test 2	Test 3	Average, %
Untreated	49.99	50.00	49.8	49.93
1% NaOH	44.39	44.45	44.17	44.34
3% NaOH	40.19	40.22	40.87	40.43
5% NaOH	30.71	30.56	30.32	30.53
7% NaOH	35.90	35.88	35.61	35.80

Conclusion

This research indicates that alkali treatment markedly improves the performance of banana fibers. Analyses using mechanical, moisture, Optical Microscope (OM), and Scanning Electron Microscopy (SEM) established that a 5% NaOH concentration is optimal for treatment. Tensile strength increased by 52%, elongation by 69%, and moisture absorption decreased by 39% at this level. Scanning electron microscopy (SEM) demonstrated the successful elimination of hemicellulose and lignin, leading to cleaner, fibrillated surfaces that enhance adhesion potential.

Elevated concentrations (7% and 9%) resulted in over-etching, the formation of cracks, and fiber fragmentation, which contributed to a reduction in strength and an increase in hydrophilicity. The elevated elongation at break signifies improved toughness and flexibility, allowing composites reinforced with 5% treated fibers to absorb greater energy prior to failure. This property is essential for applications requiring impact resistance and moisture sensitivity.

The treatment with 5% NaOH provides an optimal balance of strength, toughness, and hydrophobicity, indicating the potential of banana fibers for applications in automotive, packaging, and protective composites. Future research should integrate SEM with FTIR and TGA to enhance the validation of chemical and thermal modifications, as well as to assess long-term durability in polymer matrices.

Acknowledgement

This research is supported by the Ministry of Higher Education which gives research grant of the Fundamental Research Grant Scheme (FRGS) under a grant number of FRGS/1/2023/STG05/UNIMAP/02/8. The authors also acknowledge the Faculty of Mechanical Engineering Technology, Universiti Malaysia Perlis (UniMAP) for the lab facilities.

References

[1] Vishnu Vardhini, K., Murugan, R., & Surjit, R. (2017). Effect of alkali and enzymatic treatments of banana fiber on properties of banana/polypropylene composites. Journal of Industrial Textiles, 47(7), 1849–1864. https://doi.org/10.1177/1528083717714479

[2] Bekele, A. E., Lemu, H. G., & Jiru, M. G. (2023). Study of the Effects of Alkali Treatment and Fiber Orientation on Mechanical Properties of Enset/Sisal Polymer Hybrid Composite. Journal of Composites Science, 7(1), 37. https://doi.org/10.3390/jcs7010037

[3] Arumugam, C., Arumugam, G. S., Ganesan, A., & Muthusamy, S. (2021). Mechanical and water absorption properties of short Banana Fiber/Unsaturated Polyester/Molecular sieves + ZNO NanoRod hybrid nanobiocomposites. ACS Omega, 6(51), 35256–35271. https://doi.org/10.1021/acsomega.1c02662

[4] Shahapurkar, K., Gebremaryam, G., Kanaginahal, G., Ramesh, S., Nik-Ghazali, N., Chenrayan, V., Soudagar, M. E. M., Fouad, Y., & Kalam, M. (2024). An Experimental Study on the Hardness, Inter Laminar Shear Strength, and Water Absorption Behavior of Habeshian Banana Fiber Reinforced Composites. Journal of Natural Fibers, 21(1). https://doi.org/10.1080/15440478.2024.2338930

[5] Sathasivam, K. V., Haris, M. R. H. M., Fuloria, S., Fuloria, N. K., Malviya, R., & Subramaniyan, V. (2021). Chemical Modification of Banana Trunk Fibers for the Production of Green Composites. Polymers, 13(12), 1943. https://doi.org/10.3390/polym13121943

[6] Verma, D., & Goh, K. L. (2021). Effect of Mercerization/Alkali Surface Treatment of Natural Fibers and Their Utilization in Polymer Composites: Mechanical and Morphological Studies. Journal of Composites Science, 5(7), 175. https://doi.org/10.3390/jcs5070175

[7] Al-Daas, A., Azmi, A. S., Ali, F. B., & Anuar, H. (2023). The Effect of Alkaline Treatment to Pseudo-Stem Banana Fibers on the Performance of Polylactic Acid/Banana Fiber Composite. Journal of Natural Fibers, 20(1). https://doi.org/10.1080/15440478.2023.2176401

[8] Parre, A., Karthikeyan, B., Balaji, A., & Udhayasankar, R. (2019). Investigation of chemical, thermal and morphological properties of untreated and NaOH treated banana fiber. Materials Today Proceedings, 22, 347–352. https://doi.org/10.1016/j.matpr.2019.06.655

[9] Wijianto, N., Ibnu, R. M. D., & Adityarini, H. (2019). Effect of NaOH concentration treatment on tensile strength, flexure strength and elasticity modulus of banana fiber reinforced polyester resin. Materials Science Forum, 961, 10–15. https://doi.org/10.4028/www.scientific.net/msf.961.10

[10] Okafor, K. J., Edelugo, S. O., Ike-Eze, I. C. E., & Chukwunwike, S. A. (2020). Effect of soaking temperature on the tensile and morphological properties of banana stem fiber-reinforced

polyester composite. Polymer Bulletin, 78(9), 5243–5254. https://doi.org/10.1007/s00289-020-03374-2

[11] Collins, T., Zhang, M., Zhuang, X., Kimani, M., Zheng, G., & Wang, Z. (2022). Banana Fiber Degumming by Alkali Treatment and Ultrasonic Methods. Journal of Natural Fibers, 19(16), 12911–12923. https://doi.org/10.1080/15440478.2022.2079581

[12] Deepak Verma & Kheng Lim Goh. (2021). Effect of mercerization/alkali surface treatment of natural fibers and their utilization in polymer composites: Mechanical and morphological studies. Journal of Composites Science, 57. https://doi.org/10.3390/jcs5070175

[13] Checol, T. T., & Sendekie, T. A. (2025). Effect of alkali treatment on physicochemical and microstructural properties of false banana fiber. Scientific Reports, 15, 10825. https://doi.org/10.1038/s41598-025-10825-1

[14] Tesfay, H., Gebreyesus, Y., & Asmare, B. (2025). Optimization of banana fiber treatment parameters for improved textile performance. Discover Mechanical Engineering, 3, 29. https://doi.org/10.1007/s43939-025-00294-8

Frontiers of Chemical and Materials Engineering - ICoFCheM 2025　　　Materials Research Forum LLC
Materials Research Proceedings 60 (2026) 34-41　　　　https://doi.org/10.21741/9781644903971-5

Microbial Consortia - Mediated Degradation of Polyurethane from Landfill Leachate

Satishwaran V Rajeswaran[1,a], Amira Farzana Samat[1,2,b], Adilah Anuar[1,2,c*],
Thevanthini Nadaraja[1], Khairunissa Syairah Ahmad Sohaimi[1],
Norhidayah Abd Aziz[1], Noor Hasyierah Mohd Salleh[1], Noor Ainee Zainol[1]

[1]Faculty of Chemical Engineering and Technology, Universiti Malaysia Perlis, 02600 Arau, Perlis, Malaysia

[2]Centre of Excellence for Frontier Material Research (CFMR), Universiti Malaysia Perlis,64-66, Blok B, Taman Pertiwi Indah, Jalan Kangar - Alor Setar, Kampung Seriab, 01000 Kangar, Perlis

[a]slashsatishwaran@gmail.com, [b]amirafarzana@unimap.edu.my, [c]adilahanuar@unimap.edu.my

Keywords: Microbial Consortia, Plastic Waste Management, Polyurethane Biodegradation

Abstract. This study investigates the biodegradation of polyurethane (PU) using microbial consortia (MC) sourced from landfill leachate, aiming to enhance degradation efficiency compared to pure cultures. It involved incubating PU samples with enriched MC for 30 days and 60 days, during which various metrics were assessed: gravimetric weight loss, cell growth, enzymatic activity, and structural changes. Results indicated that microbial cell counts in the MSM+PU+MC cultures increased significantly over time, with a weight loss of 2.96 % after 60 days, surpassing the results from the 30 days cultures. FTIR analysis confirmed substantial degradation of urethane and ester bonds in PU after 60 days in the presence of MC. SEM images revealed significant surface breakdown and weakening of the polymer structure due to prolonged incubation. TGA shows that MC did promote partial degradation, primarily affecting the weaker segments of the polymer. The hard segments of PU, however, largely remained intact and degraded only at higher temperatures. Overall, these findings demonstrate that landfill leachate-derived microbial consortia possess significant potential for PU biodegradation, offering a promising foundation for developing sustainable strategies to mitigate plastic waste accumulation.

Introduction

Polyurethane (PU) is widely used due to its durability and versatility, but its resistance to degradation poses serious environmental concerns. Despite their utility, the extensive use and poor end-of-life management of PU products have raised significant environmental concerns. Conventional disposal methods like landfilling and incineration release harmful by-products and microplastics, while recycling remains energy-intensive and inefficient. Given these constraints, microbial biodegradation of PU has emerged as a promising eco-friendly alternative. Recent studies have explored landfill environments as rich sources of xenobiotic-degrading microbes, including both bacterial and fungal species with specialized metabolic pathways adapted to toxic substrates [1]. Landfill leachate and groundwater have been shown to contain diverse microbial communities, some of which produce extracellular enzymes such as hydrolases and oxidases that capable of breaking down complex polyurethane structures into simpler, non-toxic intermediates [2]. This study investigates PU film degradation using MC isolated from landfill leachate, leveraging their synergistic enzyme activity. Compared to pure cultures, MC showed greater efficiency, evaluated through weight loss, microbial growth, and structural analysis using FTIR, SEM, and TGA over 30 and 60 days. Findings highlight MC as a promising eco-friendly approach to PU waste management.

Frontiers of Chemical and Materials Engineering - ICoFCheM 2025   Materials Research Forum LLC
Materials Research Proceedings 60 (2026) 34-41   https://doi.org/10.21741/9781644903971-5

Methodology

Sample Collection and Preparation

Leachate samples were collected from Tapak Pelupusan Sanitari Rimba Emas, Chuping Perlis, Malaysia using a sterile 1000 mL bottle at 40 cm depth. Serial dilutions (10^{-1} to 10^{-4}) were performed, and 50 µL from each dilution was inoculated onto culture media to isolate bacteria and fungi [3].

Polyurethane (PU) Pre-Treatment

1 g of PU foam was cut into small pieces, disinfected with 70 % ethanol, dried, and weighed. The samples were UV-treated (UVB-15, 254 nm) for 24 hours to initiate photodegradation, making the polymer more susceptible to microbial breakdown [2].

Media Preparation

Potato dextrose agar (PDA) and nutrient agar (NA) were prepared by dissolving the pre-mix in 1 L water, boiling, and autoclaving at 121 °C for 15 minutes. After cooling to 45–50 °C, the media were poured into petri dishes, solidified, and pre-incubated at 35 ± 2 °C for 24 hours to ensure sterility before use.

Identification of Microbial Consortia in Leachate Samples

(a) Gram and Lactophenol Blue Staining Bacterial smears were prepared on glass slides, heat-fixed, and stained using Gram staining to differentiate Gram-positive and Gram-negative bacteria. Crystal violet and Gram's iodine were applied, followed by decolorization and counterstaining [13] to enable morphological observation. Lactophenol blue dye was also used to visualize microbial structures [11].

(b) Genetic Identification Genomic DNA was extracted using the PrimeWay Leachate DNA Extraction Kit and quality assessed by 1% TAE agarose gel electrophoresis, spectrophotometry (Implen NanoPhotometer® N60/N50), and fluorometric quantification. The 16S rRNA V3–V4 region was amplified with locus-specific primers containing Illumina overhang adapters. Amplicons were indexed with the Nextera XT Index Kit v2 and library quality was validated using the Agilent Bioanalyzer 2100 and Helixyte Green™ quantification. Normalized libraries were pooled and sequenced on the Illumina MiSeq platform (2×300 bp paired-end).

Incubation Method

MC were first cultured in MSM-PU wood varnish at 30 °C, 220 rpm for 7 days [4]. After 7 days, 2 mL of culture was transferred into 25 mL of fresh MSM-PU varnish and incubated for another 7 days. To conserve the community, 500 µL glycerol from the original enrichment was used to inoculate new cultures. Two experimental batches were prepared using erlenmeyer flasks containing PU samples and MC, incubated at 30 °C and 200 rpm. The first batch was withdrawn after 30 days, and the second batch after 60 days. PU samples were rinsed with sterile water and oven-dried at 50 °C until a constant weight was reached [2].

Gravimetric Analysis and Microbial Cell Count

PU samples were recovered via sterile filtration, washed with 70 % ethanol, and oven-dried at 50 °C overnight before final weighing to determine weight loss [2]. Microbial cell growth was monitored by measuring optical density at 600 nm using a spectrophotometer [3].

Fourier Transform Infrared Spectroscopy

FTIR-ATR spectroscopy was used to detect changes in the chemical structure of PU. Spectral scans were recorded in the frequency range of 4000–450 cm^{-1}, targeting specific functional groups such as urethane (C=O, N–H, C–O) to observe bond scission due to microbial action. A reduction or shift in carbonyl stretching bands indicates cleavage of urethane bonds, suggesting microbial degradation [2].

Scanning Electron Morphology

The surface morphology of degraded PU was examined using scanning electron microscopy. PU samples were sputter-coated with a palladium layer at a current of 18 mA under an argon atmosphere at 150 kPa to prevent charging. Micrographs were captured at a maximum magnification of 200 times, revealing physical erosion, pitting, and surface cracking which indicates microbial attack on the polymer structure over time.

Thermogravimetric Analysis

The analysis assessed the thermal stability of PU before and after biodegradation. Samples were heated from 30 °C to 600–700 °C at a constant rate of 10 °C/min under both inert (nitrogen) and oxidative (synthetic air) environments. The method monitors the weight loss of the polymer as a function of temperature, revealing decomposition stages, residual mass, and onset of degradation. The derivative thermogravimetry (DTG) curves were used to determine T_max, the temperature at which the maximum rate of degradation occurs [5].

Result & Discussion

Staining Analysis of Bacteria and Fungi

Figure 1(a) show chains of cocci [7]. As the staining was gram-positive, the genus that may present are *Staphylococcus, Streptococcccus, Enterococcus* and *Micrococcus.* Figure 1(b) displays cocci in pairs, known as diplococci [8]. The common genera for Gram-negative bacteria likely are *Neisseria, Moraxella, and Veionella* characterized by a thin peptidoglycan layer and inability to retain crystal violet [6]. Figures 1 (c) reveal the presence of *Fungal mycelium*, indicating active fungal growth through extensive hyphal networks. These are the possible organisms that may presented in a larger volume in the microbial community obtained from landfill.

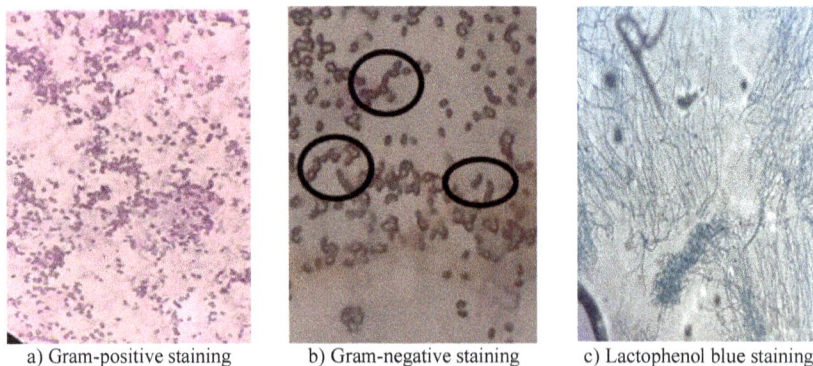

| a) Gram-positive staining | b) Gram-negative staining | c) Lactophenol blue staining |

Figure 1: Staining result. (a) Gram-positive staining. (b) Gram-negative staining. (c) Lactophenol blue staining

Gravimetric Analysis

The results showed in Figure 2 indicates the culture of microbial consortia (MC). when MC were added (MSM+PU+MC), the degradation of PU was more effective. After 30 days, the sample with microbes lost 4.27 % of its weight, and by day 60, the weight loss reached 7.23 %. This shows that the microbes helped break down the PU faster and more efficiently. Even though the samples were still in big segments, the results indicates that microbes are able to degrade polyurethanes.

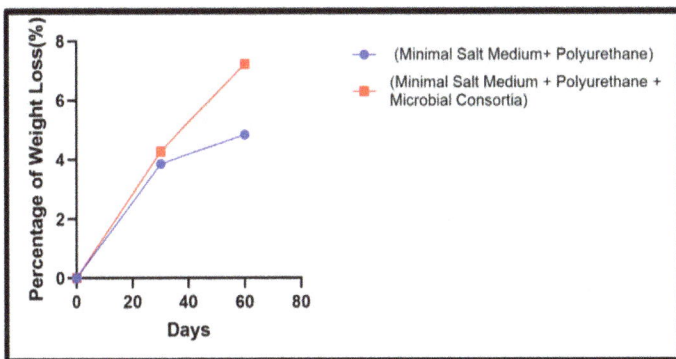

Figure 2: Gravimetric analysis

Optical Density Analysis

Figure 3 presents the optical density data in different cultures measured before incubation, and after 30 days and 60 days. In the MSM+PU culture, initial microbial presence was low, as the broth remained clear. MSM+PU+MC (with microbial consortia) had higher cell density, with gradual growth by 30 days and more significant growth by 60 days as the colour broth turns cloudy. This indicates MC started utilizing PU as a carbon source over time. For MSM+MC (without PU), absorbance was initially high but decreased over time. The lack of PU caused the microbes to depend on minimal salts and existing organic residues, leading to reduced viability. In the potato dextrose broth culture, the cell density was moderate on day 0, dropped significantly on 30 days (likely due to lag phase stress), and slightly recovered by 60 days. Manna et al. (2024) noted that toxic by-products during early metabolism can lower cell viability. In the nutrient broth culture, the density increased substantially by 30 days due to the nutrient-rich medium, but slightly declined by 60 days possibly due to nutrient depletion and natural cell death. Comparing MSM+PU and MSM+PU+MC confirms that MC improve growth, though the effect was modest. Nutrient-rich broth like potato dextrose broth and nutrient broth supported consistent microbial growth over time, whereas minimal salt media without PU resulted in poor microbial proliferation.

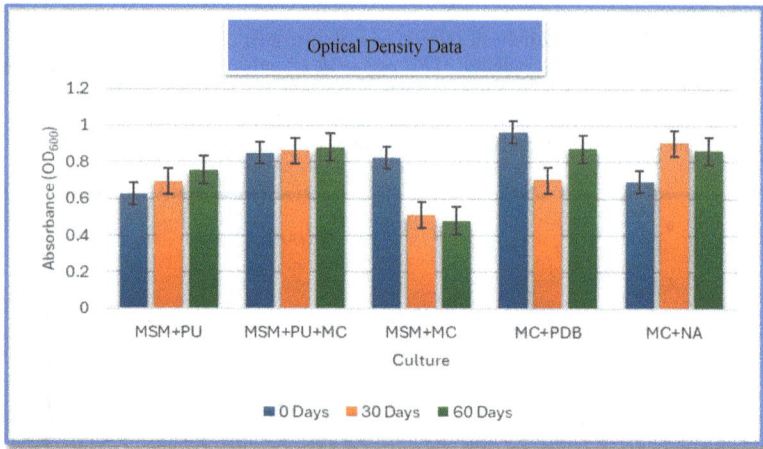

Figure 3: Optical Density Analysis

Fourier Transform Infrared Spectroscopy (FTIR)

Structural degradation of PU exposed to MC over 30 days and 60 days were evaluated. Spectra were recorded in the range of 4000–450 cm^{-1} to observe the difference in structure caused by microbes. PU samples incubated with MC (MSM+PU+MC) for 30 days exhibited large pronounced degradation. Peak reductions at 3300 cm^{-1} (urethane bonds) and 1720 cm^{-1} (ester bonds), as well as broadening in the 1100 cm^{-1} region, pointed to the formation of oxidized degradation products such as alcohols and acids. After 60 days, more extensive polymer breakdown was recorded in the MSM+PU+MC. The near disappearance of the 3300 cm^{-1} peak and significant decline at 1700 cm^{-1} indicated advanced urethane and ester bond cleavage. The broad signals between 1000–1200 cm^{-1} confirmed the accumulation of oxidized compounds such as ketones, alcohols, and carboxylic acids.

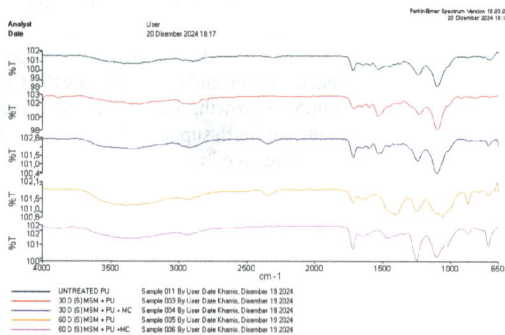

Figure 4: FTIR analysis

Scanning Electron Morphology

Secondary electrons and backscattered electrons are among the signals produced by the electron beam's interaction with the sample. These signals are used to produce fine-grained pictures of the sample's composition and topography. Figure 5 (a) shows the results of PU specimens that were obtained from SEM analysis for the untreated PU sample that appears to be smooth and uniform with no visible degradation. The structural changes after 30 days, Figure 5 (b) of incubation with the microbial consortia, noticeable surface alterations were observed, including pits, cracks, and irregular textures. After 60 days, Figure 5 (c), the PU exhibited more pronounced degradation, with deeper pits, extensive cracking, and thinning of the polymer branches, which appeared close to complete structural disintegration. These findings indicate that prolonged incubation allows the MC to accelerate the breakdown process [9]. Compared to single microbial colonies, MC demonstrate enhanced degradation efficiency by producing enzymes.

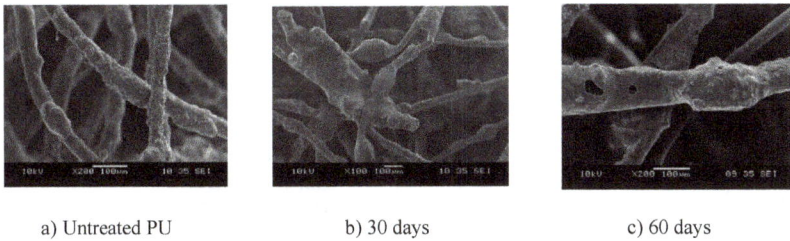

| a) Untreated PU | b) 30 days | c) 60 days |

Figure 5: SEM analysis. (a) Untreated PU. (b) 30 days MSM+MC+PU. (c) 60 days MSM+MC+PU

Thermogravimetric Analysis (TGA)

Thermal stability of PU before and after microbial incubation. The untreated PU showed high stability, with degradation starting at 358.25 °C and retaining 98.29 % of its weight, indicating an intact polymer structure. 60 days of incubation in minimal salt medium, the onset of degradation shifted to a higher temperature at 414.97 °C with 88.17 % of weight remaining. This slight increase suggests the formation of intermediate degradation products, possibly forming a temporary protective barrier. TGA revealed two distinct degradation phases, corresponding to the breakdown of PU's soft and hard segments [3]. Although the onset temperature rose, a steeper curve in the 60 days sample indicated significant structural breakdown, particularly of the hard segments.

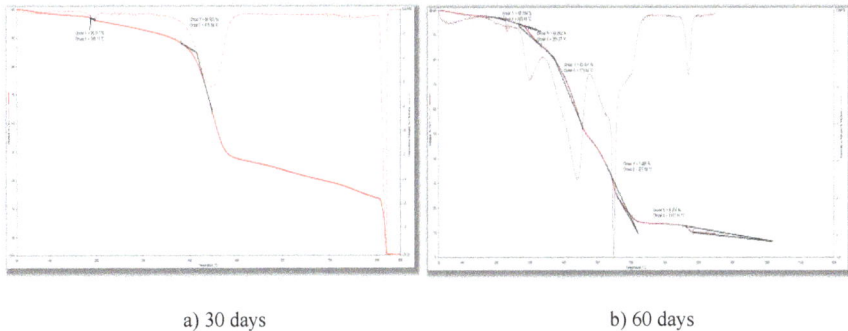

| a) 30 days | b) 60 days |

Figure 6: TGA analysis. (a) 30 days MSM+MC+PU. (b) 60 days MSM+MC+PU

Summary

This study explores the biodegradation potential of MC isolated from landfill leachate in breaking down PU. Leachate samples were analyzed through Gram and lactophenol blue staining, followed by genetic identification. Dominant microbial genera such as *Burkholderiaceae, Trupoeraceae, Acaligenaceae, Dothideomycetes,* and *Pertusaria amara* were identified, many of which produced urease and esterase enzymes. Gravimetric weight loss analysis over 60 days demonstrated that PU degradation was significantly enhanced in the presence of MC, while spectrophotometric readings confirmed microbial growth using PU as a carbon source. FTIR analysis indicated functional bond cleavage, SEM revealed surface erosion and fragmentation, and TGA showed reduced thermal stability together confirming progressive PU breakdown. These findings confirm that MC offer degradation efficiency over single strains due to their synergistic enzyme activity and adaptability. Overall, the study supports the potential use of landfill-derived MC as a solution for PU waste management.

Acknowledgement

This research was supported by the 6th Research Grant of the International Polyurethane Technology Foundation.

References

[1] Garcete, L. A. A., Martinez, J. E. R., Barrera, D. B. V., Bonugli-Santos, R. C., & Passarini, M.R. Z. (2022). Biotechnological potential of microorganisms from landfill leachate: isolation, antibiotic resistance and leachate discoloration. Anais Da Academia Brasileira de Ciencias, 94(3). https://doi.org/10.1590/0001-3765202220210642

[2] Samat, A. F., Carter, D., & Abbas, A. (2023). Biodeterioration of pre-treated polypropylene by Aspergillus terreus and Engyodontium album. Npj Materials Degradation, 7(1). https://doi.org/10.1038/s41529-023-00342-9

[3] Su, T., Zhang, T., Liu, P., Bian, J., Zheng, Y., Yuan, Y., Li, Q., Liang, Q., & Qi, Q. (2023).Biodegradation of polyurethane by the microbial consortia enriched from landfill. Applied Microbiology and Biotechnology, 107(5–6), 1983–1995. https://doi.org/10.1007/s00253-023-12418-2

[4] Gaytán, I., Sánchez-Reyes, A., Burelo, M., Vargas-Suárez, M., Liachko, I., Press, M., Sullivan, S., Cruz-Gómez, M. J., & Loza-Tavera, H. (2020). Degradation of Recalcitrant Polyurethane and Xenobiotic Additives by a Selected Landfill Microbial Community and Its Biodegradative Potential Revealed by Proximity Ligation-Based Metagenomic Analysis. Frontiers in Microbiology, 10. https://doi.org/10.3389/fmicb.2019.02986

Liu, J., Zeng, Q., Lei, H., Xin, K., Xu, A., Wei, R., Li, D., Zhou, J., Dong, W., & Jiang, M. (2023). Biodegradation of polyester polyurethane by Cladosporium sp. P7: Evaluating its degradation capacity and metabolic pathways. Journal of Hazardous Materials, 448. https://doi.org/10.1016/j.jhazmat.2023.130776

[6] Świderek, K., Martí, S., Arafet, K., & Moliner, V. (2024). Computational study of the mechanism of a polyurethane esterase A (PueA) from Pseudomonas chlororaphis. Faraday Discussions. https://doi.org/10.1039/d4fd00022f

[7] Rajabimashhadi, Z., Naghizadeh, R., Zolriasatein, A., Bagheri, S., Mele, C., & Esposito Corcione, C. (2023). Hydrophobic, Mechanical, and Physical Properties of Polyurethane Nanocomposite: Synergistic Impact of Mg(OH)2 and SiO2. Polymers, 15(8). https://doi.org/10.3390/polym15081916

[8] Cao, Z., Yan, W., Ding, M., & Yuan, Y. (2022). Construction of microbial consortia for microbial degradation of complex compounds. In Frontiers in Bioengineering and Biotechnology (Vol. 10). Frontiers Media S.A. https://doi.org/10.3389/fbioe.2022.1051233

[9] Burelo, M., Gaytán, I., Loza-Tavera, H., Cruz-Morales, J. A., Zárate-Saldaña, D., Cruz-Gómez,M. J., & Gutiérrez, S. (2022). Synthesis, characterization and biodegradation studies of polyurethanes: Effect of unsaturation on biodegradability. Chemosphere, 307. https://doi.org/10.1016/j.chemosphere.2022.136136

[10] Ranganathan, V., & Akhila, C. (2019). Streptococcus mutans: has it become prime perpetrator for oral manifestations? Journal of Microbiology & Experimentation, 7(4). https://doi.org/10.15406/jmen.2019.07.00261

[11] Hardy. (2020). Lactophenol cotton blue stain intended use. www.cms.hhs.gov/clia.

[12] Li, H., Li, L., Chi, Y., Tian, Q., Zhou, T., Han, C., Zhu, Y., & Zhou, Y. (2020). Development of a standardized Gram stain procedure for bacteria and inflammatory cells using an automated staining instrument. MicrobiologyOpen, 9(9). https://doi.org/10.1002/mbo3.1099

Frontiers of Chemical and Materials Engineering - ICoFCheM 2025
Materials Research Proceedings 60 (2026) 42-50

Materials Research Forum LLC
https://doi.org/10.21741/9781644903971-6

Cryogenic Treatment and Its Effect on The Microstructure and Corrosion Behavior for Properties Enhancement of D3 Tool Steel for Blade Application

Nur Maizatul Shima Adzali[1,3,a] *, Zuraidawani Che Daud[2,3,b],
Nur Hidayah Ahmad Zaidi[1,3,c], Siti Emy Elysha Mohd Daud[1], and
Tuan Ummi Alifah Tuan Ali[1]

[1]Faculty of Chemical Engineering and Technology, Universiti Malaysia Perlis, Kompleks Pusat Pengajian Taman Muhibbah, 02600, Jejawi, Arau, Perlis, Malaysia

[2]Faculty of Mechanical Engineering and Technology, Universiti Malaysia Perlis, Kampus Pauh Putra, 02600, Pauh, Arau, Perlis, Malaysia

[3]Centre of Excellence Frontier Materials Research, Universiti Malaysia Perlis, Perlis, Malaysia

[a]shima@unimap.edu.my, [b]zuraidawani@unimap.edu.my, [c]hidayah@unimap.edu.my

Keywords: Corrosion, Cryogenic, D3 Tool Steel, Microstructure

Abstract. This study investigates the microstructural characteristics, properties, and corrosion behavior of D3 tool steel, focusing on the effects of cryogenic treatment for blade applications. The study involved three batches of samples: untreated, heat-treated with cryogenic treatment, and heat-treated without cryogenic treatment. Untreated samples were controlled samples while heat-treated samples were austenitized at 1000°C for 30 minutes, oil quenched at 980°C for 20 minutes, and tempered at 150°C and 280°C for 90 minutes, followed by air cooling. A subset of heat-treated samples underwent additional cryogenic treatment by immersion in liquid nitrogen at -196°C for 24 hours. Comprehensive analysis using X-ray diffraction (XRD), scanning electron microscopy (SEM), and Rockwell Hardness testing was conducted before and after a corrosion test in 3.5% NaCl. The corrosion rate was measured using the weight loss method. The microstructural analysis revealed that untreated D3 steel consisted of pearlite, ferrite, and irregular carbide particles. Quenching transformed retained austenite into martensite, and cryogenic treatment refined the microstructure further, enhancing hardness. Tempering after cryogenic treatment reduced hardness but improved ductility. Heat-treated D3 steel tempered at 150°C exhibited the highest hardness (53.6 HRC). Corrosion tests showed that untreated, heat-treated with cryogenic and tempered at 150°C, and heat-treated with cryogenic and tempered at 280°C samples had lower corrosion rates compared to both heat-treated samples. Ultimately, cryogenic treated samples tempered at 280°C (HTC280) demonstrated optimal characteristics for blade applications, and balancing hardness (50.2 HRC) with superior corrosion resistance for prolonged performance in corrosive environments.

Introduction

Tool steels used for blade applications are carbon and alloy steels specifically designed for cutting instruments such as knives and industrial blades. The specific type of tool steel chosen for blade applications depends on the desired properties, such as hardness, wear resistance, and response to heat treatment. The AISI D series of tool steels, known for their high carbon and chromium content, including D2, D3, D5, D6, and D7, are commonly used for making press tools, knives, and blades due to their wear resistance [1]. AISI D3, in particular, is noted for its high abrasion and wear resistance, and high compressive strength which often used in the industrial sector for dies and cold-work instruments, and its composition includes 2-2.35% carbon and 11-13% chromium [2]. The high chromium content in D3 steel provides good corrosion resistance by

Frontiers of Chemical and Materials Engineering - ICoFCheM 2025   Materials Research Forum LLC
Materials Research Proceedings 60 (2026) 42-50   https://doi.org/10.21741/9781644903971-6

forming a protective oxide layer on the surface. Cryogenic treatment involves slowly cooling material from room temperature to cryogenic temperatures around -196°C [3]. This process enhances mechanical characteristics by converting residual austenite to martensite and precipitating carbides within the lattice [1]. Research by Kumar et al. (2019) indicated that D3 specimens subjected to 12 hours of cryogenic treatment showed noticeable increases in microhardness and tensile strength [4].

The D3 tool steel samples received in annealed state have soft and ductile with low tensile strength. Batheja et al. (2012) noted, annealed steel comprises ferrite, pearlite, and carbides, which are unsuitable for blade applications [2]. Therefore, heat treatment is necessary to transform the microstructure and improve hardness. Additionally, corrosion can weaken the blade and impair cutting capacity, making it essential to conduct corrosion tests in a 3.5% sodium chloride solution to understand the corrosion behavior of treated and untreated D3 tool steel samples.

Methodology
Experimental Procedures
Annealed D3 tool steel (as-received) was provided and cut to a size of 10 mm x 10 mm and 20 mm x 20 mm. To get rid of surface oxide and contaminants, these samples were grounded, polished and etched with Nital solution before microstructure analysis. Table 1 shows a group of D3 tool samples that underwent a different heat treatment process.

Table 1: Treatment process for D3 tool steel

Nomenclature	Heat Treatment
UT	Untreated
HT150	Heat-treated and tempered at 150°C
HT280	Heat-treated and tempered at 280°C
HTC150	Heat-treated with cryogenic and tempered at 150°C
HTC280	Heat-treated with cryogenic and tempered at 280°C

The D3 samples underwent austenitizing heat treatment at 1000°C, oil quenching at 980°C, cryogenic treatment for 24 hours in liquid nitrogen at -196°C, and tempering at two different temperatures, either 150°C or 280°C. One batch of these steels remain untreated as a controlled sample. The other batch then divided into another two batches which both of this batches underwent heat treatment but only one batch will undergo cryogenic treatment. Figure 1 shows the overall heating profile for D3 tool steel. The samples have undergone a corrosion immersion test to study the effect of corrosion behavior on untreated and treated samples of D3 tool steel. The samples were immersed in 3.5% NaCl for 30 days thus, the weight loss of the sample is recorded every three days. The corrosion rate of D3 tool steels was calculated by using Eq. 1.

$$\text{Corrosion rates, } C_R = \frac{KW}{AT\rho}\left(\frac{mm}{yr}\right) \tag{1}$$

Microstructure analysis was performed using the JEOL JSM 6460 Scanning Electron Microscope (SEM) to observe the microstructure of the samples. X-ray diffraction (XRD) analysis using Bruker D2 Phase used to determine the phase structure present. Rockwell hardness testing was performed according to ASTME18-16 with 120° diamond indenter and 150kgf (1471N) of total force. All of these tests and analysis were done before and after the corrosion test to observe changes occur to the mechanical and corrosion properties of D3 tool steel.

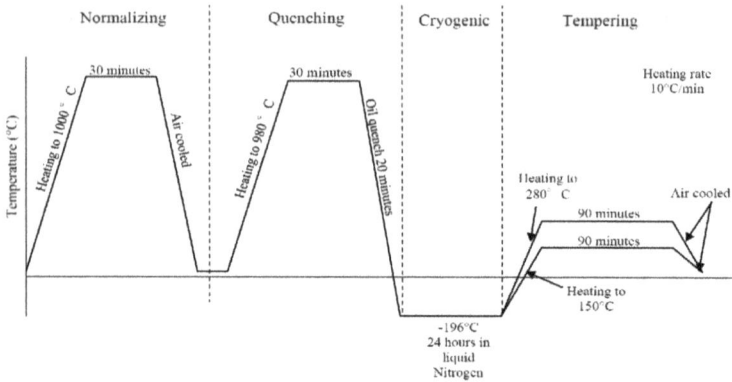

Fig. 1: Overall heating profile for D3 tool steel

Result and Discussion

Microstructure analysis before corrosion test

Figures 2(a)-(e) revealed the microstructures of the D3 tool steel with different heat treatment. Figure 2(a) shows the untreated (UT) sample in annealed condition which is mainly consisting of pearlite, chromium primary carbide and ferrite indicating a soft and ductile condition [2]. Figures 2(b)-(c) shows the microstructures of D3 that austenitized at 1000°C, oil quenching at 980°C and tempering at temperature 150°C (HT150) and 280°C (HT280) respectively. When austenization converts ferrite and pearlite phases to austenite phases, another transformation from austenite to martensite occurs during quenching. As a result, the microstructure of D3 when quenched in oil, changes to martensite, retained austenite, and undissolved primary chrome carbide. D3 appears to have formation of martensite and chromium carbides after quenching process that leads to the increment of hardness [5]. This tempered martensite was formed after tempering that causes reduction of hardness, which will be discussed further in Table 2.

Figures 2(d)-(e) shows the microstructure of D3 that austenitized, oil quenching, cryogenic treatment and tempering at temperature 150°C (HTC150) and 280°C (HTC280) respectively. As the quenching process causes the production of martensite and retained austenite, cryogenic treatment is performed right after oil quenching to promote the transformation of retain austenite and martensite phases to full martensite phases by exposing the sample to an extremely low temperature (-196°C). Higher tempering temperatures result in little to no retained austenite and demonstrate the early formation of fine secondary carbides, especially with less primary carbide compared to lower tempering temperatures [6]. The presence of secondary carbide can increase the mechanical properties such as hardness and strength.

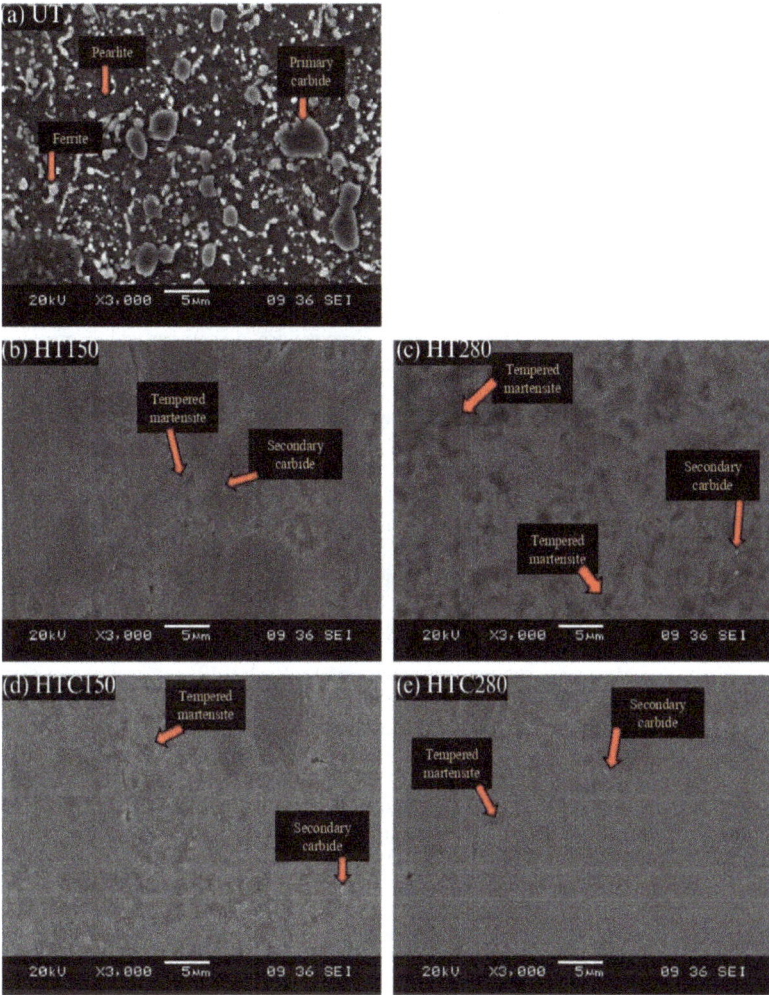

Fig 2: SEM images of D3 sample for (a) UT, (b) HT150, (c) HT280, (d) HTC150, (e) HTC280.

Hardness of D3 tool steel before and after corrosion test

The hardness of D3 sample for UT, HT and HTC presented in Table 2. According to the Table 2, it shows that the UT sample has the lowest reading of hardness before and after the corrosion test. HT and HTC samples both have the approximately comparable. HT150 has the highest hardness (53.6 HRC) before corrosion, and after corrosion test (52.9 HRC).

45

Frontiers of Chemical and Materials Engineering - ICoFCheM 2025   Materials Research Forum LLC
Materials Research Proceedings 60 (2026) 42-50   https://doi.org/10.21741/9781644903971-6

Table 2: Hardness of D3 tool steels before and after corrosion test

D3 sample	Hardness before corrosion test (HRC)	Hardness after corrosion test (HRC)
UT	14.4	11.9
HT150	53.6	52.9
HT280	52.9	52.2
HTC150	50.4	48.7
HTC280	50.2	49.5

The untreated (UT) D3 steel has the lowest hardness due to formation of ferrite, carbide and pearlite phase that present in the microstructure and make it soft and ductile [2]. HT and HTC gained the increment of hardness after receiving the heat treatment from normalizing, oil quench, tempered including cryogenic treatment. Cryogenic treatment is an add-on process that improves the mechanical properties by transforming retained austenite to martensite and initiating the nucleation site, precipitating a large number of fine carbides in the matric martensite [9]. Tempering temperature also plays an important role in enhancement of the D3 tool steel hardness. The increasing tempering temperature can cause the decreasing of hardness due to decomposition of high carbon martensite to low carbon martensite and carbide, transformation of retained austenite to ferrite and carbide [6]. During corrosion test, the aggressive interaction of chloride ions (Cl^-) with the steel surface, which speeds up the corrosion process, is the main reason for the weight loss of steel in NaCl solution [5]. When steel is submerged in NaCl solution, the chloride ions compete with the dissolved oxygen and hydroxide ions for adsorption on the metal [5]. Adsorption disturbs the steel's protective oxide layer, resulting in the production of iron chloride compounds. Furthermore, the presence of dissolved oxygen from the air promotes iron oxidation, resulting in the development of iron oxides and hydroxides (rust). This combination of chemical processes causes iron ions to dissolve and be removed from the steel, resulting in significant weight loss. As a result, the removal of material from the steel surface, combined with the production of porous and less dense corrosion products, weakens the overall surface structure and reduces hardness.

Phase analysis before and after corrosion test

X-ray diffraction (XRD) results for phase analysis before and after corrosion test for D3 are represented in Figure 3. Based on the result, high intensity peaks were observed at ~45°, with smaller peaks at ~65° and ~82°. In Figure 3(a) peak 110 represents the martensite pattern that shows minor signs of retained austenite in the sample [4]. This conversion of phases from retained austenite to martensite phase is responsible for improvement of mechanical properties as the result of fine microstructures [4]. Relatively weak peaks that are observed in all XRD patterns for all samples represent retained austenite [7]. In Figure 3(b) corrosion occurs on the D3 samples affecting the intensity of peak. High peak intensity indicates high number of atoms in the crystal lattice which also indicate it has high degree of crystallinity or larger crystal size. For the UT sample, the peak intensity at 45° decreased after the corrosion test. It shows that the chemical reaction occurs due to the reaction of the surface and the corrosive solution has weakened the surface structure. The decrease of the intensity of peak indicates the decrease of the degree of crystallinity of the sample which is related to the decrease of the hardness. This is because, when amorphous regions are softer, less dense, and allow more chain/atomic mobility. These zones deform more easily, which will reducing overall hardness. As for the HT and HTC sample, the intensities of the peak are slightly different. This is also related with hardness of the sample before and after corrosion test where it shows a slight decrease of hardness.

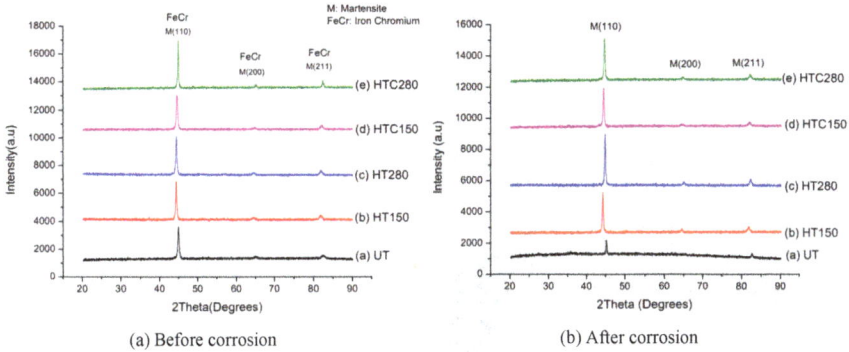

(a) Before corrosion   (b) After corrosion

Fig 3: XRD pattern for sample D3 tool steel (a) before and (b) after corrosion immersion test.

Microstructure analysis after corrosion test

Corroded surface of D3 samples represented in Figure 4. Based on the observation after 30 days of immersing the samples in 3.5% NaCl solution, it shows presence of rust on the surface of the samples as a result of reaction between D3 samples and NaCl solution. It also can be observed that pitting corrosion forms on the D3 samples. Figures 4(b)-(c) which represent samples HT150 and HT280 are more prone to pitting corrosion and this is proved by the result for corrosion rate in Table 3 which stated that those samples have higher corrosion rate.

Based on the result calculated for the corrosion rate of D3 tool steel (discussed in Table 3) corrosion rate for HT150 and HT280 have the highest corrosion rate compared to UT, HTC150 and HTC280, which microstructure shown in Figures 4(a), 4(d) and 4(e). This result shows that D3 tools steel that received heat treatment without cryogenic treatment (HT samples) easily deteriorate when reacting to a corrosive environment which in this case is 3.5% NaCl solution. Deep cryogenic treatment alters microstructure by increasing carbide density and significantly reducing retained austenite content [8]. Although the increased carbide density could enhance corrosion resistance, the drastic reduction in austenite leads to a decrease in corrosion resistance [8]. In this study, HT has higher hardness than HTC which indicates that it is possible to exhibit low corrosion resistance because of the reduction of austenite phase and high carbide density in the microstructure. Higher hardness can sometimes improve corrosion resistance because a harder surface is more resistant to scratches, wear, and plastic deformation. Fewer surface defects can lead to fewer sites for corrosive attack to initiate (e.g., pitting or crevice corrosion).

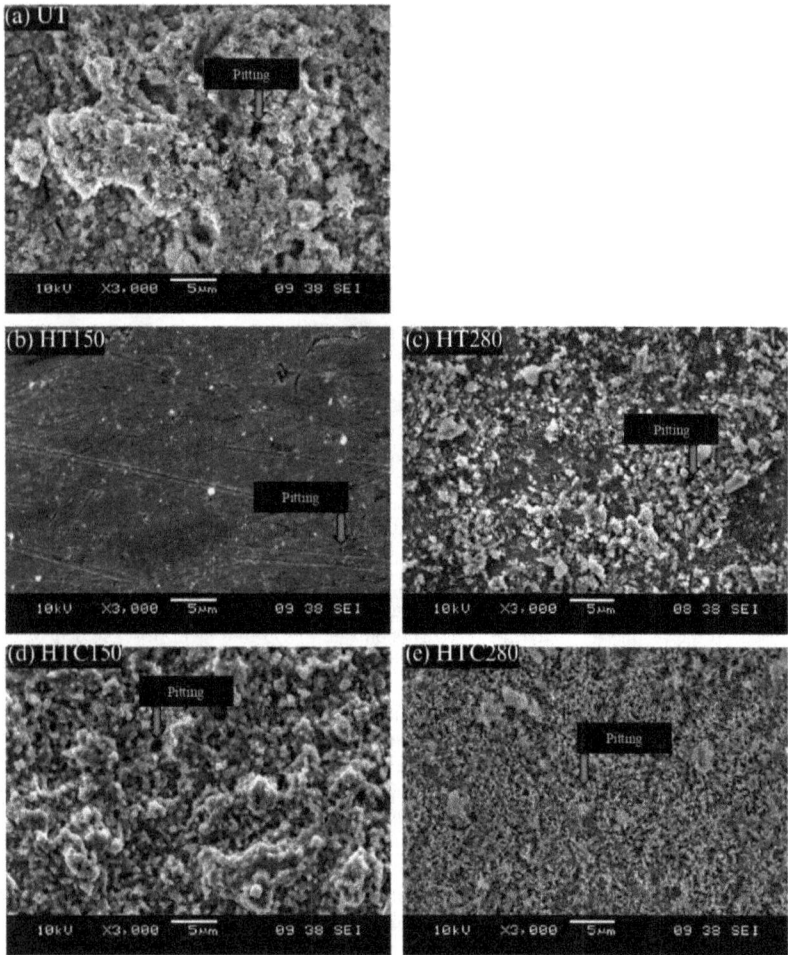

Fig 4: SEM images showing microstructure of D3 tool steel for (a) UT, (b) HT150, (c) HT280, (d) HTC150 and (e) HTC 280 after corrosion immersion test in 3.5% NaCl solution.

Corrosion behavior of D3 tool steel

Table 3 provides calculated data of corrosion rate in millimeters per year (mm/year) for all the D3 samples which calculated using an Eq. (1). UT has the lowest corrosion rate, attributed to its pearlite–carbide–ferrite microstructure, which is ductile and soft with low hardness. Thus, low carbide formation can be the reason the corrosion rate is low because high carbide can act as galvanic cells that accelerate corrosion rate [10]. By referring to the Table 2, hardness for the HTC sample is lower than HT sample due to low formation of homogeneous carbide. Thus, HTC also produces high formation of solutionised chromium atoms in structure after receiving cryogenic treatment which can be the reason it has a low corrosion rate. Chromium plays a crucial role in enhancing the corrosion resistance of steel as it produces a protective layer and prevents galvanic corrosion due electrical contact in the presence of electrolyte by presence of two dissimilar metals [10]. Decrease free chromium atoms and increased formation of fine carbide making more grain boundaries as the potential sites for the increasing of the corrosion rate [10].

Table 3: Corrosion rate of D3 tool steel

D3 sample	Corrosion rate (mm/year)
UT	0.3817
HT150	0.5725
HT280	0.5725
HTC150	0.3817
HTC280	0.3817

Summary

As a conclusion for this research, HT150 shows the highest hardness among other samples. In terms of corrosion behavior, samples UT, HTC150 and HTC280 seem to have the lowest corrosion rate value. For the purpose of blade application, heat treated with cryogenic treatment, tempered at 280°C (HTC280) will be chosen because it has comparable hardness (50.2 HRC) with HT150 (53.6 HRC). Thus, the HTC280 has good corrosion resistance that helps to maintain the performance and longer lifespan even if it is used in a corrosive environment.

References

[1] S. Saad Ghazi and K. Mijbel Mashloosh, Influence of Heat Treatment on Resistance of Wear and Mechanical Properties of Die Steel Kind D3, (2015). doi.org:10.5251/ajsir.2015.6.2.33.40.

[2] A. Bhateja, A. Varma, A. Kashyap, and B. Singh, Study the Effect on the Hardness of Three Sample Grades of Tool Steel i.e., and D3 after Heat Treatment Processes Such as Annealing, Normalizing, and Hardening & Tempering (2012) 253-259. ISSN: 2319-1813 ISBN: 2319-1805.

[3] A. Wale and V. Wakchaure, Effect of Cryogenic Treatment on Mechanical Properties of Cold Work Tool Steels (2013) 149-154. ISSN: 2249-6645.

[4] S. Kumar, M. Nagaraj, A. Bongale, and N. K. Khedkar, Effect of deep cryogenic treatment on the mechanical properties of AISI D3 tool steel (2019). DOI: 10.1504/IJMATEI.2019.099789 https://doi.org/10.1504/IJMATEI.2019.099789

[5] M. May, Corrosion behavior of mild steel immersed in different concentrations of NaCl, Journal of Sebha University-(Pure and Applied Sciences), Vol.15 No.1 (2016). 1-12.

[6] M. A. Mochtar, W. N. Putra, and M. Abram, Effect of tempering temperature and subzero treatment on microstructures, retained austenite, Mater Res Express, vol. 10, no. 5 (2023). https://doi.org/10.1088/2053-1591/acd61b

[7] N. W. Khun, E. Liu, A. W. Y. Tan, Effects of deep cryogenic treatment on mechanical and tribological properties of AISI D3 tool steel, Friction, vol. 3 (3) (2015) 234-242. https://doi.org/10.1007/s40544-015-0089-z

[8] P. Jovičević-Klug, T. Kranjec, M. Jovičević-Klug, Influence of the deep cryogenic treatment on AISI 52100 and AISI D3 steel's corrosion resistance, Materials, vol. 14, no. 21 (2021) 6357. https://doi.org/10.3390/ma14216357

[9] J. X. Zou, T. Grosdidier, B. Bolle, K. M. Zhang, and C. Dong, Texture and microstructure at the surface of an AISI D2 steel treated by high current pulsed electron beam, Metall Mater Trans A Phys Metall Mater Sci, vol. 38 A (9) (2007) 2061-2071. doi.org: 10.1007/s11661-007-9146-1. https://doi.org/10.1007/s11661-007-9146-1

[10] K. Amini, A. Akhbarizadeh & S. Javadpour. Effect of Carbide Distribution on Corrosion Behavior of the Deep Cryogenically Treated Steel, J Mater Eng Perform, vol. 25(2) (2016) 365-373. https://doi.org/10.1007/s11665-015-1858-6

Frontiers of Chemical and Materials Engineering - ICoFCheM 2025
Materials Research Proceedings 60 (2026) 51-56

Materials Research Forum LLC
https://doi.org/10.21741/9781644903971-7

The Influence of Sintering Temperature of Powder Metallurgy Mg-3wt.% Zn/6wt.% β-TCP: The Hardness and Corrosion Behavior

Zuraidawani Che Daud[1,2,a] *, Mohd Nazree Derman[1,2,b],
Muhammad Asri Mohd Salleh[1,c], Muhamed Yazid Farhan Rohazad[1,d] and
Nur Maizatul Shima Adzali[2,3,e]

[1]Faculty of Mechanical Engineering & Technology, Universiti Malaysia Perlis, Kampus Pauh Putra, 02600 Arau, Perlis, Malaysia

[2]Center of Excellence for Frontier Materials Research, Universiti Malaysia Perlis, 02600 Arau, Perlis, Malaysia

[3]Faculty of Chemical Engineering & Technology, Universiti Malaysia Perlis, Kompleks Pusat Pengajian Taman Muhibbah, 02600, Jejawi, Arau, Perlis, Malaysia.

[a]zuraidawani@unimap.edu.my, [b]nazree@unimap.edu.my,
[c]s211351933@studentmail.unimap.edu.my, [d]yazidfarhan81@gmail.com,
[e]shima@unimap.edu.my

Keywords: Beta Tricalcium Phosphate, Magnesium-Zinc, Corrosion, Hardness, Powder Metallurgy

Abstract. Powder Metallurgy (PM) is an advanced manufacturing technique that enables the fabrication of materials with unique microstructures and properties. Magnesium (Mg) alloys, particularly magnesium–zinc (Mg-Zn) alloys, are gaining popularity in biomedical applications due to their lightweight nature, good strength, and compatibility with biological systems. However, their mechanical and corrosion properties still need improvement, which can be achieved through alloying and reinforcement. This study investigates the effect of different sintering temperatures on the hardness and corrosion behavior of Mg-3wt.% Zn/6wt.% β-Tricalcium Phosphate (β-TCP) composites. Mg-3wt.% Zn was mixed with 6wt.% β-TCP powder using a roll mill at 120 rpm for 60 minutes. The mixture was then compacted at a pressure of 200 MPa. Sintering was conducted in an argon-controlled atmosphere at temperatures of 450°C, 500°C, and 550°C. The microstructure of the sintered samples was examined using an optical microscope, and Vickers microhardness testing was used to measure hardness. The corrosion behavior was evaluated using a potentiostat. Results showed that sintering at 450°C produced the highest hardness and good corrosion resistance.

Introduction

The magnesium (Mg)-based alloy has garnered significant attention as a bone implant compared to other traditional metal implants, like Co-based alloys, stainless steel, and titanium. Mg-based alloys have good biocompatibility, biodegradability, and low-stress shielding effect, and are also the lightest structural alloys [1-3]. However, the clinical application of Mg-based implants faces significant challenges, primarily related to their rapid corrosion rate in physiological environments [4]. This accelerated degradation can lead to premature loss of mechanical integrity and the production of hydrogen gas, which may cause tissue damage and impair healing processes.

Zinc (Zn) has emerged as one of the most promising alloying elements for biomedical magnesium applications due to its beneficial effects on both mechanical properties and biocompatibility [5]. The presence of 3 wt.%Zn in the Mg matrix leads to the formation of intermetallic phases, which contribute to the strengthening of the alloy.

β-tricalcium phosphate (β-TCP) is a bioactive ceramic that is used for bone repair. It possesses unique characteristics that promote the growth of new bone tissue, distinguishing it from other materials used for bone grafts. The resorbable characteristic of this material closely simulates the solubility of bone materials, allowing it to integrate with new bone development [6]. Incorporating β-TCP particles in the Mg-Zn is well-acknowledged to impart valuable bioactivity properties, as β-TCP possesses excellent biocompatibility and bioactivity because of its close resemblance to the chemical and structural properties of human bone [7]. The mechanism of β-TCP with the formation of an apatite layer can improve the bioactivity performance of Mg-Zn alloy after immersion in a simulated body fluid solution.

Sintering temperature is one of the most important processing parameters in powder metallurgy, heavily affecting the microstructure, mechanical properties, and corrosion behavior of magnesium-based composites. Generally, higher sintering temperatures improve densification and particle bonding, resulting in better mechanical properties [8]. However, excessive temperatures can cause negative effects such as element segregation, especially the loss of volatile elements like zinc, coarsening of reinforcement particles, and the formation of brittle intermetallic networks [7,8]. The porosity, distribution of second phases, and interfacial characteristics developed during sintering directly influence corrosion rates and the uniformity of degradation in environments similar to physiological conditions. Therefore, optimizing the sintering temperature is essential to balance mechanical performance and corrosion resistance.

Methodology

Mg-3wt.% Zn/6wt.% β-TCP composites were prepared by powder metallurgy technique. Raw materials of Mg powder (from Sigma Aldrich) and 3wt.% of Zn powder (from Hmbg) were mixed with 6wt.% of β-TCP powders (from Sigma Aldrich). The raw materials were mixed for 60 minutes using a roll mill machine at 120 rpm. Then the mixed powders were uniaxial compacted using a hand press machine with a pressure of 150 MPa. The green compact samples were then sintered in a tube furnace at sintering temperatures of 450°C, 500°C, and 550°C, respectively, for 2 hours inside an argon atmosphere. The sintered samples were polished according to the standard preparation of the metallographic samples. For microstructural analysis, all samples were observed under an optical microscope, and the hardness values were measured using the Vickers hardness test.

NOVA software and AUTOLAB PGSTAT 204 were both implemented for the linear polarization test. A three-electrode cell system was used in the electrochemical measurements, consisting of the sintered sample as the working electrode (WE), a platinum rod as the counter electrode (CE), and a saturated calomel electrode (SCE) as the reference electrode as shown in Fig.1. To ensure a consistent exposed area, the specimen's surface was covered with electroplating tape, leaving only a 1 cm^2 region exposed. Phosphate buffer saline (PBS) solution was used as the electrolyte. The polarization scan was performed with a scan potential range from -250 mV to +250 mV at a scanning rate of 1 mVs^{-1}. The corrosion rate of Mg-3wt.% Zn/6wt.% β-TCP composites were measured using equation (1).

$$Corrosion\ Rate\ (mm/year) = 0.1288 \times I_{corr} \times \frac{EW}{\rho} \tag{1}$$

Where, EW is equivalent to the corroding species and ρ is the density in g/cm^3. The I_{corr} is the corrosion current density in unit of mAcm^{-2}.

Frontiers of Chemical and Materials Engineering - ICoFCheM 2025   Materials Research Forum LLC
Materials Research Proceedings 60 (2026) 51-56   https://doi.org/10.21741/9781644903971-7

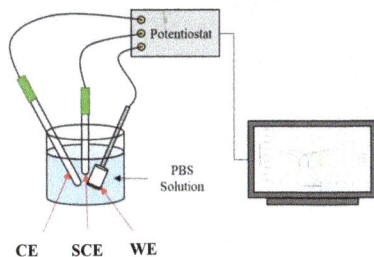

Fig. 1: The three-electrode cell setup for polarization test.
CE: *Counter Electrode,* **SCE**: *Saturated Calomel Electrode,* **WE**: *Working Electrode*

Result and Discussion

Fig. 2 shows the hardness value after sintering at three sintering temperatures 450°C, 500 °C, and 550°C, respectively. It can be seen that the trend of hardness value decreases after sintering at 500°C and then slightly increases at a sintering temperature of 550°C. From Fig. 2, sample sintered at 450°C represents the highest value of hardness, which is 36.33 HV. Meanwhile, the lowest hardness value is 24.43 HV at a sintering temperature of 500°C, and the value of hardness slightly increases to 30.05 HV at a sintering temperature of 550°C. The highest value of hardness is due to strong interparticle bonding. The β-TCP disperses through the Mg-3wt.% Zn matrix and enhances their hardness value [10]. However, at 500°C of sintering temperature, the grain coarsening of the sample gives the lowest values of hardness.

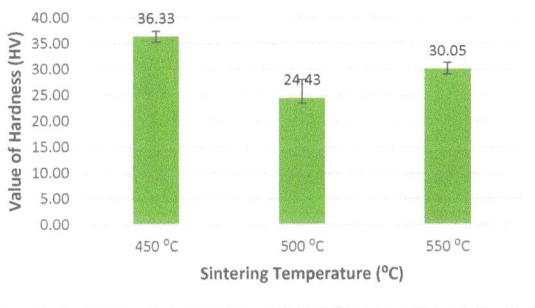

Fig.2: The value of Hardness with various sintering temperatures

The optical microstructures of Mg-3wt.% Zn/6wt.% β-TCP composites after sintering at 450°C, 500°C, and 550°C are shown in Fig. 3. The microstructure of the sample sintered at 450°C [Fig. 3a] exhibits relatively fine, well-defined grains and small pores present in low to moderate amounts. At the sintering temperature of 500°C [Fig. 3b], the grain size increases compared to the sample sintered at 450°C. Meanwhile, at 550°C [Fig. 3c], the microstructure displays coarse grains with pores located at and along the grain boundaries. The formation of pores decreases with increasing sintering temperatures.

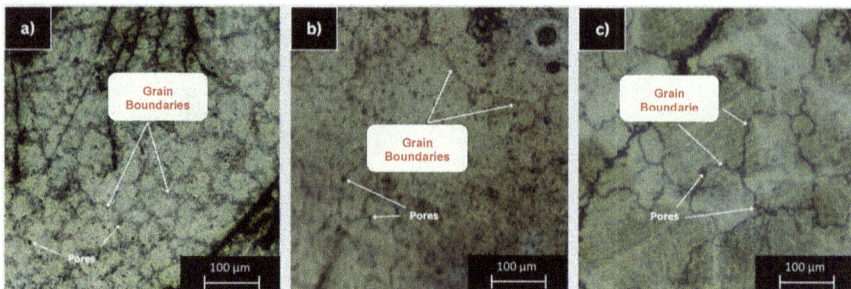

Fig.3 Optical microstructures of sample after sintered at different sintering temperatures (a) 450°C, (b) 500°C, (c) 550°C

The results of the linear polarization test performed on specimens immersed in phosphate-buffered saline (PBS) solution are presented in Table 1. The corrosion potential (E_{corr}) and corrosion current (I_{corr}) were determined by extrapolating the polarization curve depicted in Fig. 4, while the corrosion rate was extracted from the NOVA software. However, equation (1) could also be used to obtain the corrosion rate of the Mg-3wt.% Zn/6wt.% β-TCP. E_{corr} serves as an indicator of the likelihood of corrosion during the test, while I_{corr} quantifies the severity of corrosion experienced by the specimens [11].

The E_{corr} value of the sample sintered at 450°C was -1.3104 V. Meanwhile, samples sintered at 500°C and 550°C show a shift to more positive E_{corr} values, measuring -1.0942 V and -0.8034 V, respectively. The change in corrosion potential (E_{corr}) toward more positive values gives an increasing corrosion rate. Thus, the sample sintered at 450°C exhibited the most favorable corrosion performance, with the lowest corrosion rate. These findings indicated that the highest corrosion rate occurred at higher sintering temperatures, which was attributed to grain coarsening and higher porosity. Yan et al. [12] highlighted that coarse grains play a significant role in causing non-uniform corrosion and increasing localized corrosion, which results in accelerated material degradation.

Table 1: Result of linear polarization of 450°C, 500°C, and 550°C samples

Sintering Temperature (°C)	E_{corr} (V)	I_{corr} (A/cm^2)	Corrosion Rate (mm/year)
450	-1.3104	6.29×10^{-5}	0.7306
500	-1.0942	1.32×10^{-4}	1.5353
550	-0.8034	1.51×10^{-4}	1.7519

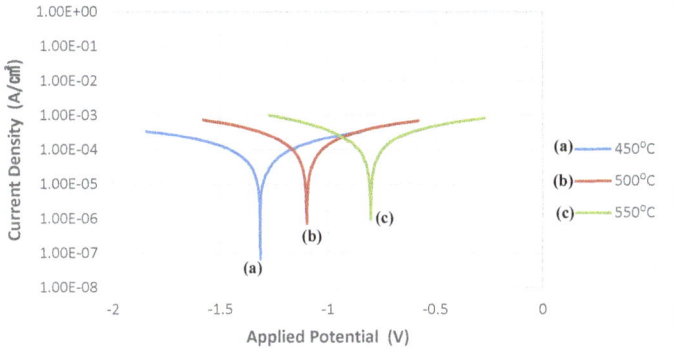

Fig.4: Tafel plot of different samples, (a) 450°C, (b) 500°C, and (c) 550°C

Fig. 5 shows the optical microstructure after corrosion testing for samples sintered at three different temperatures. It can be seen that sample sintered at 450°C (Fig.5 (a)) demonstrates minimal pitting and localized corrosion, as fine grains provide a more uniform protective layer. Meanwhile, samples sintered at 500°C have more pitting, and localized corrosion occurs due to larger grain boundaries acting as corrosion initiation sites (Fig.5 (b)). More pitting and localized corrosion occur as coarse grains create weak spots in the protective layer (Fig.5 (b)).

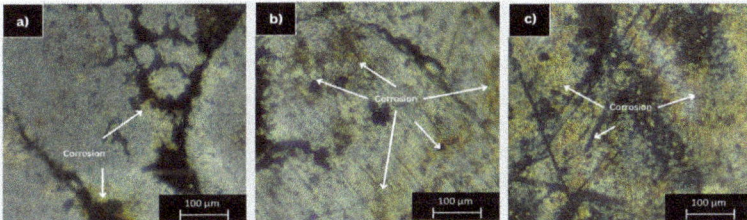

Fig.5: Optical microstructures after corrosion testing for samples sintered at different sintering temperatures, (a) 450°C, (b) 500°C, (c) 550°C

Conclusion

In conclusion, optimizing sintering temperature is essential to improve the hardness and corrosion resistance of Mg-3wt.% Zn/6wt.% β-TCP composites, making them more viable for biomedical applications. The sample sintered at 450 °C exhibited the best overall performance, with the lowest corrosion rate (0.7306 mm/year) and the highest hardness (36.33 HV). Increasing the sintering temperature to 500°C and 550°C led to higher corrosion rates and reduced hardness.

References

[1] A. C. Bîrcă, A. A. Neacşu, O. R. Vasile, I. Ciucă , I. M. Vasile, M. A. Fayeq, B. S. Vasile, Mg-Zn alloys, most suitable for biomedical applications, Rom J. Morphol Embryol. 59, (2018), 49-54. PMID: 29940611.

[2] G. Parande, V. Manakari, H. Gupta, M. Gupta, Magnesium-β-Tricalcium Phosphate Composites as a Potential Orthopedic Implant: A Mechanical/Damping/Immersion Perspective. Metals, 8, (2018), 343. https://doi.org/10.3390/met8050343

[3] K. Kumar, A. Das and S. B. Prasad, Recent developments in biodegradable magnesium matrix composites for orthopaedic applications: A review based on biodegradability, mechanical and biocompatibility perspective, Materials Today: Proceedings, 44, (2020) 2038-2042. https://doi.org/10.1016/j.matpr.2020.12.133

[4] S. K. Sharma, S. Gajević, L. K. Sharma, D. G. Mohan, Y. Sharma, M. Radojković, and B. Stojanović, Significance of the Powder Metallurgy Approach and Its Processing Parameters on the Mechanical Behavior of Magnesium-Based Materials, Nanomaterials. (2025). https://doi.org/10.3390/nano15020092

[5] Y. Huang, D. B. Liu, M. Xia, and L. Anguiliano, Characterization of an Mg-2Zn-1Ca 1β-TCP Composite Fabricated by High Shear Solidification and ECAE, Materials Science Forum. (2013). https://doi.org/10.4028/www.scientific.net/msf.765.813

[6] M. Bohner, B. L. G Santoni and N. Döbelin, Tricalcium phosphate for bone substitution: Synthesis and properties. Acta Biomaterialia. 113 (2020) 23-41. https://doi.org/10.1016/J.ACTBIO.2020.06.022

[7] J. Jeong, J. H. Kim, J. H. Shim, N. S. Hwang, and C. Y. Heo, Bioactive calcium phosphate materials and applications in bone regeneration. Biomaterials research. (2019). https://doi.org/10.1186/s40824-018-0149-3

[8] K. Narita, E. Kobayashi, S. Tatsuo. Sintering Behavior and Mechanical Properties of Magnesium/β-Tricalcium Phosphate Composites Sintered by Spark Plasma Sintering, MATERIALS TRANSACTIONS, (2016). 57, 1620-1627. https://doi.org/10.2320/matertrans.L-M2016827

[10] S. A. Buyong, S. B. Jamaludin, and R. Abdul Malek, Effect of beta tricalcium phosphate (β-TCP) on properties of Mg-Zn composites. Key Eng. Mater. (2013) 203–206. https://doi.org/10.4028/www.scientific.net/kem.594-595.203

[11] N. S. Azmi, M. N. Derman, and Z. C. Daud, AC and DC Anodization on the electrochemical properties of SS304L: A comparison. Advances in Materials Research. (2024) https://doi.org/10.12989/amr.2024.13.3.153

[12] Y. Yan, X. Chu, X. Luo, X. Xu, Y. Zhang, Y. L. Dai, D. Li, L. Chen, T. Xiao, and K. Yu, A homogenous microstructural Mg-based matrix model for orthopedic application with generating uniform and smooth corrosion product layer in Ringer's solution: Study on biodegradable behavior of Mg-Zn alloys prepared by powder metallurgy as a case. Journal of Magnesium and Alloys (2021). 9(1), 225–240. https://doi.org/10.1016/J.JMA.2020.03.010

Frontiers of Chemical and Materials Engineering - ICoFCheM 2025   Materials Research Forum LLC
Materials Research Proceedings 60 (2026) 57-63   https://doi.org/10.21741/9781644903971-8

Anodised Fe₂O₃/ SS304L as an Electrocatalyst for Hydrogen Production

Mohd Nazree Derman[1,2,a] *, Zuraidawani Che Daud[1,2,b] and Nur Suhaily Azmi[1,c]

[1]Faculty of Mechanical Engineering & Technology, Universiti Malaysia Perlis, Pauh Putra Main Campus, 026000 Arau, Perlis, Malaysia

[2]Centre of Excellence for Frontier Materials Research, Universiti Malaysia Perlis, No. 64-66, Blok B, Taman Pertiwi Indah, Jalan Kangar - Alor Setar, Kampung Seriap, 01000 Kangar, Perlis, Malaysia

[a]nazree@unimap.edu.my, [b]zuraidawani@unimap.edu.my, [c]suhaily93@gmail.com

Keywords: Anodizing, Stainless Steel, Iron Oxide Hydrogen Production and Catalysts

Abstract. This research focused on the synthesis and study the iron oxide formation on stainless steel 304L (SS304L) by anodizing method for hydrogen production. The study is carried out in AC anodizing, optimization of AC anodizing parameters such as such as anodizing voltage, concentration of ammonium fluoride (NH₄F), anodizing time, and the temperature of the electrolyte solution by using L9 Taguchi orthogonal arrays with including ANOM analysis and repetition of the experiment for the optimum parameters and lastly is the hydrogen application. The anodizing electrolyte used are ethylene glycol solution with ammonium fluoride and distilled water. Surface morphology of the anodized films were characterized using field emission scanning electron microscopy (FESEM) and transmission electron microscopy (TEM). The optimization via the Taguchi method identified the best combination of parameters such as 30 V of anodizing voltage, 0.3 M of NH₄F concentration, 15 °C of electrolyte temperature and 20 minutes of anodizing time, enhancing the electrocatalytic activity for the hydrogen evolution reaction (HER). Under optimal conditions, the AC anodized SS304L exhibited significantly improved performance, with increased stability and efficiency in water splitting applications, compared to as-received specimen. These findings suggest that by carefully controlling the anodizing parameters, it is possible to produce AC anodized SS304L with superior electrocatalytic properties, making it a viable and cost-effective option for renewable hydrogen production.

Introduction

The oxygen evolution reaction (OER) and the hydrogen evolution reaction (HER), two separate half processes, are involved in electrochemical water splitting, an innovative technique for producing high-purity hydrogen [1]. Because of the sluggish kinetics of both HER and OER, there is a need to achieve superior efficiency in electrochemical water splitting [2]. A catalyst could be used in order to overcome the problems.

Commercial catalysts like ruthenium/iridium dioxide for OER and platinum for HER are used to a certain extent because of their high prices and scarcity, which highlights the need for bifunctional, highly active, and stable catalysts [3]. As a result, there has been increased interest in transition metal compounds (TMCs) [4], notably transition metal nitrides (TMNs), which have improved catalytic activity and are hence promising options for water splitting. These compounds may have their electrical structure, morphology, and active sites changed, which opens possibilities for changing their catalytic performance [5], emphasizing the need for careful consideration in the development of catalysts. Stainless steel (SS), the most prevalent and economical alloy composed of Ni, Fe, and Cr, has gained popularity as an alternative material for energy storage and electrocatalysis because of its outstanding conductivity and durability [6]. However, the scarcity of active sites and the surface overpotential have prevented its direct application as a water splitting catalyst. Controlling SS's composition and electrical structure has helped enhance its catalytic activity.

For instance, Anantharaj and colleagues [7] enhanced electrocatalytic performance by hydrothermally eliminating Cr from the surface of 316 SS. Introducing AC anodizing as a technique for surface modification provides a practical solution to improve the HER characteristics of SS304L stainless steel. During AC anodizing, controlled modifications to the oxide film's surface morphology, composition, and structure may cause increased electrocatalytic activity. This approach offers customized enhancements in catalytic performance, which represents important progress towards realizing SS304L's full potential for hydrogen evolution.

Methodology

The SS304L(17.5–19.5% chromium, 8.0–10.5% nickel, and a maximum of 0.03% carbon) samples with a thickness of 2 mm were cut into sizes of 25 mm x 15 mm and underwent sample preparation of grinding, polishing and austenitizing at a temperature of 700 °C with a heating rate of 2 °C per minute. The sample surfaces were then covered with electroplating tape to ensure that just 10 mm × 10 mm are exposed during the anodizing process. In this research, the optimization of AC anodizing parameters applied to the L9 Taguchi orthogonal arrays.

Anodizing is a method used to create an oxide film on the surface of metal alloy shown in Figure 1. In this study, a basic anodizing setup with two electrodes (anode and cathode) is used. The SS304L is set as both anode and cathode and an AC power supply as the power source. The anodizing process is carried out using voltage ranging from 10 V to 50 V in an ethylene glycol solution comprising 0.3 M ammonium fluoride, NH_4F, and 3% water, H_2O. Using anodizing voltage lower than 10 V resulted in prolong anodizing time, which more than 30 minutes. This may because of thickness of SS304L used, which is 2 mm. The temperature of the electrolyte is fixed at room temperature, and the process is carried out for 30 minutes.

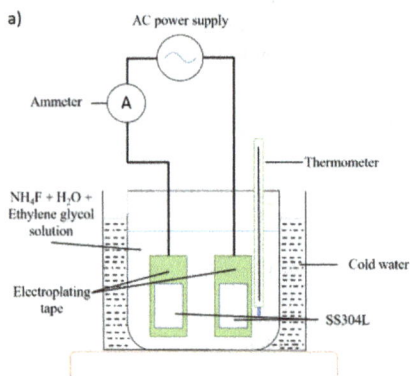

Figure 1: AC anodizing setup

The optimization of process parameters in the Taguchi method depends significantly on the SNR as a statistical tool. In this research, the signal-to-noise ratio (SNR) was calculated using the overpotential values obtained from the Linear Sweep Voltammetry (LSV) curves of HER and OER reactions at current density -0.01 Acm^{-2} and 0.01 Acm^{-2}, respectively shown in Figure 2. The L9 Taguchi orthogonal array was applied and four parameters at three different levels each were included, where the anodizing parameters were anodizing voltage (30 V, 35 V, 40 V), NH_4F

Frontiers of Chemical and Materials Engineering - ICoFCheM 2025   Materials Research Forum LLC
Materials Research Proceedings 60 (2026) 57-63   https://doi.org/10.21741/9781644903971-8

concentration (0.2 M, 0.3 M, 0.4 M), electrolyte temperature (15 °C, 25 °C, 35 °C) and anodizing time (20 min, 30 min, 40 min). The AC anodized specimens were analyzed on the morphological and electrocatalytic activity performance observations using the field emission scanning electron microscope (FESEM) model Hitachi SU8030, transmission electron microscope (TEM) and selected area electron diffraction (SAED) analysis were conducted using the FEI Tecnai G2 F20 X-TWIN model.

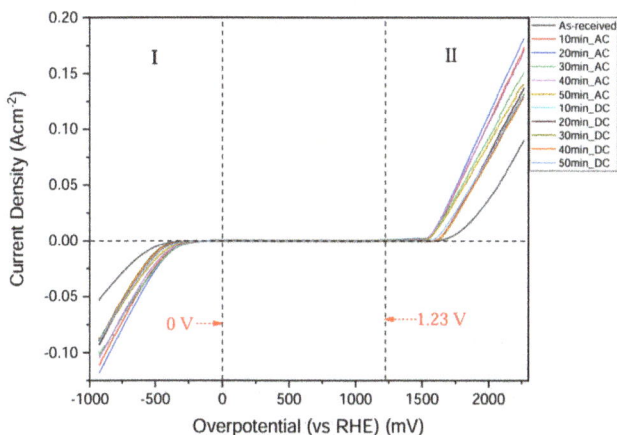

Figure 2. Overall linear sweep voltammetry of this experiment.

Results and Discussion

Table 1 displayed the response table of η_{HER} and the SNR values for each parameter with the factor level. The analysis revealed the optimal combination for a lower η_{HER} was obtained at 30 V of anodizing voltage (level 1), 0.3 M of NH_4F concentration (level 2), 15°C of electrolyte temperature (level 1) and 20 minutes of anodizing time (level 1). An optimal combination was selected according to the smaller-the better (STB) objective function, where the lowest SNR values were chosen for each parameter. The four parameters were ranked based on the value of delta obtained, which was calculated by subtracting the highest value obtained in the response table (Table 2), with the lowest value for each factor. The anodizing voltage was considered the most influential parameter, as rank 1, followed by concentration of NH_4F, electrolyte temperature and anodizing time. This finding was aligned with the research by Wang et al. [8] which studied the effect of different anodizing parameters on the SS304. The research concluded that the different anodizing voltages had affected the oxide film formed the most compared to the other anodizing parameters. The oxide film was one of the crucial factors affecting the electrocatalytic activity performance of the electrocatalyst. These optimal values were interpreted on a combined graph and labelled with a red circle, as shown in Figure 3.

Table 1: Result of HER overpotential value, η, in linear sweep voltammetry (LSV) test with the mean SNR result.

No.	Voltage (V)	NH₄F (M)	Temperature (°C)	time (min)	Noise Factor Speed 1	Noise Factor Speed 2	Signal-to-noise ratio, η_{HER} (dB)
					Control Factors		HER Overpotential value, ŋ (mV)
1	30	0.2	15	20	398.30	393.42	-51.9510
2	30	0.3	25	30	378.77	390.98	-51.7075
3	30	0.4	35	40	347.03	388.54	-51.3257
4	35	0.2	15	20	366.56	347.03	-51.0516
5	35	0.3	25	30	376.33	364.12	-51.3705
6	35	0.4	35	40	371.45	359.24	-51.2553
7	40	0.2	15	20	364.12	361.68	-51.1958
8	40	0.3	25	30	390.98	378.77	-51.7075
9	40	0.4	35	40	354.36	361.68	-51.0786

The SNR plot as depicted in Figure 3 was important as the plot showed the trend of the factors or parameters with the optimum level and which of the factors influence the output response. A main effect was presented when the levels of a factor affect the output response differently. If the graph line was parallel to the x-axis, then there is no main effect presented. The greater differences in vertical position, the greater the magnitude of the main effect. Based on the SNR plot in Figure 3, the red circle indicates the optimum level for lower HER overpotential value which were 30 V anodizing voltage, 0.3 M of NH₄F concentration, 15 °C of electrolyte temperature and 20 minutes of anodizing time. A repetition test by applying the optimum level of anodizing parameter's combination would improve the HER overpotential value and increase the electrocatalytic activity performance of the AC anodized SS304L as the electrocatalyst for water splitting application

Table 2:Response table for HER overpotential, ηHER SNR where the * indicates the lowest SNR values for each factor

Level	Voltage	Concentration	Temperature	Time
1	-51.66*	-51.40	-51.64*	-51.47*
2	-51.23	-51.60*	-51.28	-51.39
3	-51.33	-51.22	-51.30	-51.36
Delta	0.44	0.38	0.36	0.11
Rank	1	2	3	4

Main Effects Plot for SN ratios
Data Means

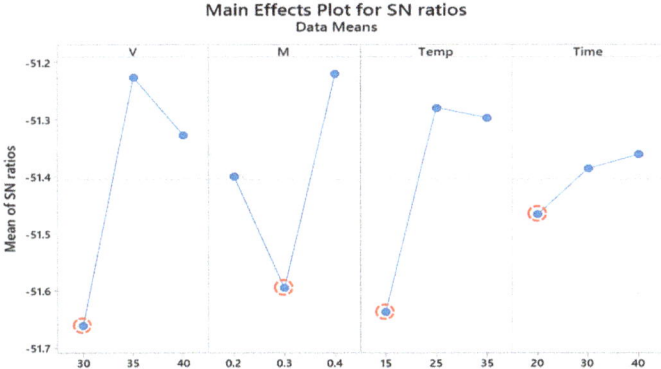

Figure 3: The SNR plot for overpotential value, ηHER, with the red circle, which denotes the optimal parameters.

For the best combination specimen, the surface morphology revealed a porous spherical-like structure and pore formation on the surface of AC anodized SS304L, as depicted in Figure 4 The spherical-like structure was clearly seen. This finding was aligned with the studies by Rahman et al. [9], who had declared the typical shape of iron oxide nanoparticles to be spherical morphology. This could enhance the electrocatalytic properties as the active surface area was also increased. As studied by Gebreslase et al. [10], a larger active surface area exhibited splendid electrocatalytic activity for both HER reactions.

Figure 4: FESEM morphological images AC anodizing parameters,

Based on Figure 5a, the TEM images of iron oxide nanoparticles revealed the state of amorphous and agglomerated nature. The agglomerated particles could be clearly observed in the dark region exhibited in the images. The amorphous materials showed a lack of long-range order because of a random atomic arrangement without a defined crystal lattice structure, as shown in Figure 5b. The SAED image shows no plane in this image (Figure 5b). This resulted in an irregular and porous morphology of the nanoparticles. This finding was aligned with the studies by Martín-González et al., [11], who had conducted anodizing process on iron foil in ethylene glycol solution and the result of the lamella TEM analysis exhibited the structure of the pore of the iron oxide film formed. The studies had concluded that the oxide film composed of quasi-amorphous structure. The pore of the oxide film formed was comprised multiple regions where the first was a layer of completely amorphous shell around the pore walls, the second was a layer mixture of amorphous and nanocrystal regions and a layer of interphase between the substrate and the oxide film.

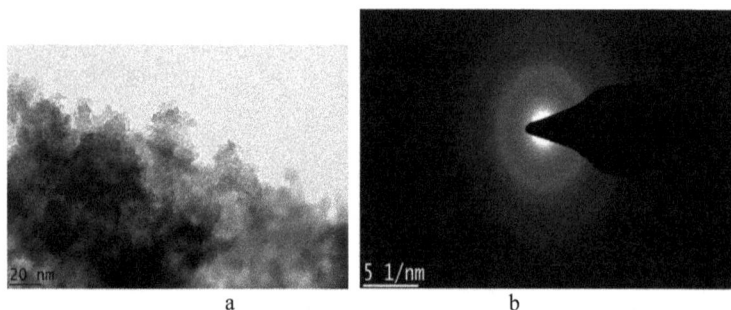

a   b

Figure 5:TEM analysis of iron oxide powder obtained from the optimum AC anodized SS304L at 130k magnification (b) SAED analysis of the specimen

Conclusion

The research investigated the many correlations between the SS304L stainless steel's HER properties, oxide film shape, and AC anodizing voltage. The optimisation of AC anodizing parameters found 30 V anodizing voltage, 0.3 M NH_4F concentration, in 15°C electrolyte temperature at 20 minutes anodizing time. An important focus of the work is how the anodizing voltage affects how the oxide film on the surface of SS304L. The TEM and SAED analyses of the iron oxide powder obtained from the optimized specimen detected the amorphous nature of the iron oxide.

Acknowledgment

The author would like to acknowledge the support from the Fundamental Research Grant Scheme (FRGS) under the grant number FRGS/1/2019/TK10/UNIMAP/02/2 from the Ministry of Higher Education Malaysia. Special thanks to Centre of Excellence for Frontier Materials Research (FrontMate) and Faculty of Mechanical Engineering and Technology for providing facilities and testing this project.

References

[1] Du, J., Xiang, D., Zhou, K., Wang, L., Yu, J., Xia, H., Zhao, L., Liu, H., & Zhou, W. (2022) Electrochemical hydrogen production coupled with oxygen evolution, organic synthesis, and waste reforming. Nano Energy, 104 (PA), 107875. https://doi.org/10.1016/j.nanoen.2022.107875

[2] Amin, M., Shah, H. H., Fareed, A.G., Khan, W. U., Chung, E., Zia, A., Rahman Farooqi, Z. U., & Lee, C. (2022) Hydrogen production through renewable and non-renewable energy processes and their impact on climate change. International Journal of Hydrogen Energy 47 (77), 33112-33134. https://doi.org/10.1016/j.ijhydene.2022.07.172

[3] Yao, M., Sun, B., Wang, N., Hu, W., & Komarneni, S. (2019) Self-generated N-doped anodized stainless steel mesh for an efficient and stable overall water splitting electrocatalyst. Applied Surface Science, 480 (January), 655-664. https://doi.org/10.1016/j.apsusc.2019.03.036

[4] Gao, Y., Xiong, T., Li, Y., Huang, Y., Li, Y., & Balogun, M. S. J. T. (2019) A Simple and Scalable Approach to Remarkably Boost the Overall Water Splitting Activity of Stainless Steel Electrocatalysts. ACS Omega. 4 (14), 16130-16138. https://doi.org/10.1021/acsomega.9b02315

[5] Su, H., Jiang, J., Song, S., An, B., Li, N., Gao, Y., & Ge, L. (2023) Recent progress on design and applications of transition metal chalcogenide-associated electrocatalysts for the overall water splitting. Chinese Journal of Catalysis, 44, 7-49. https://doi.org/10.1016/S1872-2067(22)64149-4

[6] Zhu, J., Hu, L., Zhao, P., Lee, L. Y. S., & Wong, K. Y. (2020) Recent Advances in Electrocatalytic Hydrogen Evolution Using Nanoparticles. Chemical Reviews, 120 (2), 851-918. https://doi.org/10.1021/acs.chemrev.9b00248

[7] Anantharaj, S., Chatterjee, S., Swaathini, K. C., Amarnath, T. S., Subhashini, E., Pattanayak, D. K., & Kundu, S. (2018) Stainless Steel Scrubber: A Cost Efficient Catalytic Electrode for Full Water Splitting in Alkaline Medium. ACS Sustainable Chemistry and Engineering, 6 (2), 2498-2509. https://doi.org/10.1021/acssuschemeng.7b03964

[8] Wang, Yingge, Li, G., Wang, K., & Chen, X. (2020b). Fabrication and formation mechanisms of ultra-thick porous anodic oxides film with controllable morphology on type-304 stainless steel. Applied Surface Science, 505, 144497. https://doi.org/10.1016/j.apsusc.2019.144497

[9] Rahman, S. S. U., Qureshi, M. T., Sultana, K., Rehman, W., Khan, M. Y., Asif, M. H., Farooq, M., & Sultana, N. (2017). Single step growth of iron oxide nanoparticles and their use as glucose biosensor. Results in Physics, 7, 4451-4456. https://doi.org/10.1016/j.rinp.2017.11.001

[10] Gebreslase, G. A., Martínez-Huerta, M. V., Sebastián, D., & Lázaro, M. J. (2023). NiCoP/CoP sponge-like structure grown on stainless steel mesh as a high performance electrocatalyst for hydrogen evolution reaction. Electrochimica Acta, 438(November 2022). https://doi.org/10.1016/j.electacta.2022.141538

[11] Martín-González, M., Martinez-Moro, R., Aguirre, M. H., Flores, E., & Caballero Calero, O. (2020). Unravelling nanoporous anodic iron oxide formation. Electrochimica Acta, 330. https://doi.org/10.1016/j.electacta.2019.135241

Frontiers of Chemical and Materials Engineering - ICoFCheM 2025   Materials Research Forum LLC
Materials Research Proceedings 60 (2026) 64-68   https://doi.org/10.21741/9781644903971-9

Effect of Temperature and Time on Polyurethane Degradation in Deep Eutectic Solvents

Amirah Nasuha Mohd Razib[1,2,a], Mohd Sharizan Md Sarip[1,2,b] *,
Nik Muhammad Azhar Nik Daud[1,c], Amirul Ridzuan Abu Bakar[1,d],
Mohd Asraf Mohd Zainudin[1,e], Siti Kartini Enche Ab Rahim[1,f] and Zuhaili Idham[3,g]

[1]Faculty of Chemical Engineering & Technology, Universiti Malaysia Perlis, 02600, Arau, Perlis, Malaysia

[2]Centre of Excellence for Frontier Materials Research (FrontMate), Universiti Malaysia Perlis, 02600, Perlis, Malaysia

[3]Department of Deputy Vice-Chancellor (Research and Innovation), Universiti Teknologi Malaysia, Johor, Skudai, 81310, Malaysia.

[a]amirahrazib@studentmail.unimap.edu.my, [b]sharizan@unimap.edu.my,
[c]nikazhar@unimap.edu.my, [d]amirulridzuan@unimap.edu.my, [e]mohdasraf@unimap.edu.my,
[f]sitikartini@unimap.edu.my, [g]zuhailiidham@utm.my

Keywords: Degradation, Polyurethane, Deep Eutectic Solvents

Abstract. This study focuses on the degradation of polyurethane (PU) using deep eutectic solvents (DES) to support more environmentally friendly methods for recycling PU waste. The DES used in this study was prepared from choline chloride and analytical grade urea, a non-toxic compound widely used in fertilizers and pharmaceuticals, which are known to have low toxicity and a good ability to dissolve many materials including biomass-derived compounds, synthetic polymers and various metal oxides or salts. This makes them suitable for chemical recycling processes. The degradation experiments were conducted at different temperatures (160°C, 170°C, and 180°C) and for various heating times (2, 4, 6, and 8 hours). The results showed that complete degradation of PU occurred at 170°C after 8 hours and at 180°C after 4 to 8 hours. These results show that higher temperature helps to break down PU more effectively. The findings suggest that choline chloride and urea can be used as effective solvents for PU degradation under heat and can help to support safer and greener chemical recycling of plastic materials.

Introduction

Polyurethane (PU) is widely used in many industries, such as foams, coatings, adhesives and elastomeric materials [1]. It is chosen for its strong mechanical strength and resistance to chemicals [2]. However, its widespread use leads to a large amount of plastic waste. PU does not degrade easily in nature and is hard to recycle through conventional methods like melting or mechanical processing [3]. Most PU waste is either landfilled or burned, causing soil pollution, toxic gas release and high carbon emissions [3], [4].This is mainly due to its strong cross-linked structure, making reuse or dissolution difficult.

Chemical recycling provides an alternative by recovering valuable monomers. Deep eutectic solvents (DES) are formed by the interaction between a hydrogen bond donor (HBD) and a hydrogen bond acceptor (HBA), are gaining interest as environmentally friendly solvents for polymer degradation due to their less volatile, biodegradable and lower toxicity than conventional organic solvents [5], [6], [7] One among the most promising of these for cleaving strong polymer bonds, including urethane linkages, is choline chloride–urea DES [8], [9]. This type of DES is considered safe and non-hazardous, consistent with the principle that chemical recycling of thermoset polymers should not produce harmful by-products. However, the use of DES in

Frontiers of Chemical and Materials Engineering - ICoFCheM 2025   Materials Research Forum LLC
Materials Research Proceedings 60 (2026) 64-68   https://doi.org/10.21741/9781644903971-9

thermoset plastic recycling, such as PU, is still at an early stage and with a limited understanding of the underlying reaction mechanisms.

This study investigates the solubility and degradation of PU in ChCl and urea DES under controlled conditions and analyzes the influence of physicochemical characteristics such as density and viscosity on degradation efficiency. This study offers enhanced understanding of the interaction between DES and PU at increased temperatures by comparing experimental results with published data and assessing degradation products. These findings facilitate the connection between fundamental solubility research and practical chemical recycling methodologies, providing direction for sustainable polymer waste management.

Methodology

Materials
The high-flow thermoplastic polyurethane (TPU) was supplied by Revlogi Materials Solutions Sdn Bhd (Malaysia). Urea and choline chloride (ChCl) were purchased from Sigma-Aldrich, Germany. The source of acetone was HMBG Chemical in Germany.

Preparation of DES
The mixture of ChCl and urea was prepared using a 1:2 molar ratio. The mixing process occurred at 70°C for 2 hours until the liquid became clear, transparent, and uniform in appearance.

Physical properties of DES
The density of DES components was measured using a pycnometer. The viscosity of the deep eutectic solvent was determined using a viscometer (Brookfield, United Kingdom).

Degradation of PU in DES
In this study, 5 g of PU and 5 g of DES were mixed in a 100 ml beaker. The degradation process was carried out at temperatures of 160 °C, 170 °C and 180 °C with different heating durations of 2, 4, 6 and 8 hours. Then, the sample underwent through a post-treatment step.

Post-treatment process after degradation of PU
The degradation product underwent post-treatment to separate unreacted PU, DES and o-toluidine. Whatman filter paper was used to filter the mixture once it had cooled to room temperature. The solid, filter cake contained unreacted PU and the liquid, filtrate 1, contained DES and o-toluidine. To separate o-toluidine which is insoluble in water from DES, distilled water was added to filter 1. This step was repeated to obtain filter cake 2 and filtrate 2. Filter cake 2 was dried at 55 °C and weighed as W1. Filtrate 2 was processed with vacuum rotary evaporation at 65 °C and 100 Pa to recover DES.

Filter cake 1 underwent washing with distilled water to remove remaining DES, which was recovered by rotary evaporation. The cleaned filter cake 1 was dissolved in acetone to extract unreacted PU. A third filtration gave filter cake 3 and filtrate 3. Filtrate 3 contained acetone and minor degradation products, while filter cake 3 contained unreacted PU. Acetone was removed from filtrate 3 by vacuum rotary evaporation at 35 °C, and the recovered PU was dried and recorded as W2. Filter cake 3 was also dried and recorded as W3. W1 was identified as o-toluidine and W3 as unreacted PU.

$$\text{Degradation rate} = \frac{W_O - W_3}{W_O} \times 100 \qquad (1)$$

Where:

W_0 = Initial weight of PU

W_3 = Weight of PU not degraded

Frontiers of Chemical and Materials Engineering - ICoFCheM 2025   Materials Research Forum LLC
Materials Research Proceedings 60 (2026) 64-68   https://doi.org/10.21741/9781644903971-9

Results and Discussion

Physical properties of DES

The measured density of the synthesized DES was found to be 1.19 g/cm³ ± 0.01, which is equivalent to the literature value of 1.24 g/cm³ [10]. Table 1 shows the physical properties of ChCl and urea in comparison with a previous study. The elevated density of this DES results from the robust hydrogen bonding interactions between ChCl and urea, resulting in a compact molecular structure [5]. These interactions reduce intermolecular distances, leading to a dense and cohesive solvent structure. The effectiveness of ChCl and urea DES in solubilizing polymers is enhanced by its strong hydrogen-bonding environment [9].

Table 1   *Properties of DES*

Properties	Value		Ref
	This Work	Literature Review	
Density, ρ (gcm^{-3})	1.19	1.24	[10]
Viscosity, μ (cP)	755	750	[5]

The viscosity of the synthesized DES was 755 cP ± 0.14, which agrees well with the literature value of 750 cP [5]. The high viscosity arises from the ionic nature of ChCl and its interaction with urea, which slows down molecular mobility and forms a structured fluid network through extensive hydrogen bonding [11]. This elevated viscosity directly influences flow behavior and interaction rates with polyurethanes.

Degradation product

The degradation efficiency of PU in the presence of DES was significantly affected by temperature and reaction time. Figure 1 shows the degradation profiles at different temperatures and time intervals. At 160°C, the degradation rate was minimal, with only 35.78% degradation observed after 2 hours and 39.10% after 8 hours. This indicates that 160°C does not provide sufficient thermal energy to overcome the activation energy barrier necessary for efficient urethane bond cleavage [12]. Increasing the temperature to 170°C markedly enhanced degradation. After two hours, about 75.55% of the PU had degraded. Extending the reaction time did not further increased degradation. After 4 and 8 hours, degradation reduced to approximately 71.21% and 56.06%, respectively. The observed reduction may result from intermediate degradation products, including stable urethane fragments or cyclic compounds, which exhibit resistance to further degradation.

Optimal degradation was observed at 180°C. Degradation reached approximately 60% within 2 hours, with complete degradation, 100% achieved by 4 hours. This suggests that elevated thermal energy improves the capacity of DES to cleave urethane linkages and effectively degrade the PU backbone [13]. The primary degradation product identified was o-toluidine, resulting from the cleavage of urethane bonds. Zhang et al. reported the complete degradation of PU under milder conditions, specifically at 170°C for 8 hours [9]. This highlights the importance of thermal activation in overcoming the energy barriers associated with the cleavage of stable chemical bonds in polyurethane.

The optimum temperature identified in this study is comparatively high, yet it is still markedly lower than that of conventional disposal methods. Incineration is a combustion process characterized by high temperature, ranging from 750 °C and 1200 °C, facilitating the rapid oxidation of waste [14]. This process converts refuse into bottom ash, heat, and flue gas or toxic emissions. In contrast to incineration, degradation in DES at 180 °C prevents the emission of harmful gases and employs a biodegradable, non-volatile solvent, thereby adhering more closely to environmentally sustainable practices. Furthermore, complete degradation in 4 hours at 180 °C is more energy-efficient than 8 hours at 170 °C, as noted by Zhang et al. [9], since the reduced duration may offset the increased heating demand. Future research should concentrate on reducing energy demand through the introduction of catalysts, the application of microwave or ultrasonic irradiation or the evaluation of alternative deep eutectic solvent formulations, thereby improving the sustainability of the process.

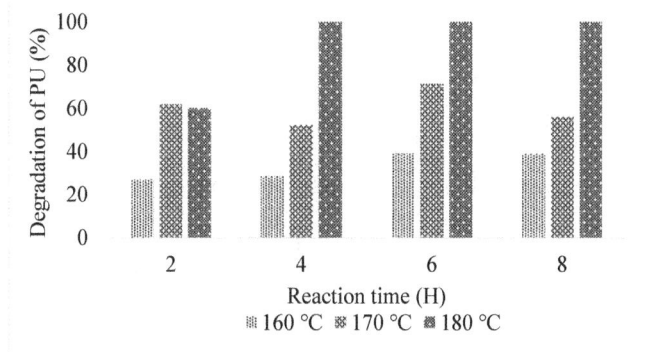

Figure 1   Degradation rate of PU at different temperatures (160, 170 and 180°C) and times (2,4,6 and 8 hours).

Conclusion

This study shows that deep eutectic solvents, particularly a combination of choline chloride and urea, efficiently promote the thermal degradation of polyurethane. The findings show that complete degradation of polyurethane can occur at elevated temperatures of 170°C for 8 hours or 180°C for 4 hours, underscoring the efficacy of deep eutectic solvents as a sustainable and effective medium for polyurethane recycling. These findings promote the advancement of eco-friendly techniques for the management of polyurethane waste, providing an alternative to traditional recycling practices. Future study must emphasize the optimization of reaction conditions and the scale up of the processes to commercial levels, thereby facilitating the advancement of sustainable polymer recycling technologies.

Acknowledgement

The financial support provided by the Ministry of Higher Education Malaysia through the Fundamental Research Grant Scheme (FRGS), reference no. FRGS/1/2023/TK05/UNIMAP/02/12, and the International Polyurethane Technology Foundation, Japan, is greatly appreciated.

References

[1] S. A. Jadhav, A. V. Rane, K. Kanny, S. S. Mulge, V. K. Abitha, and S. Thomas, *Application of Blends and Polyurethane Interpenetrating Polymer Networks*. J. Polyurethane Polymers, 2017, pp. 359-375. https://doi.org/10.1016/B978-0-12-804039-3.00015-4

[2] D. Wienen, T. Gries, S. L. Cooper, and D. E. Heath, An overview of polyurethane biomaterials and their use in drug delivery, *J. Control. Release*, 2023, pp. 376–388. https:doi.org/10.1016/j.jconrel.2023.09.036

[3] M. B. Johansen, B. S. Donslund, S. K. Kristensen, A. T. Lindhardt, and T. Skrydstrup, Tert-Amyl Alcohol-Mediated Deconstruction of Polyurethane for Polyol and Aniline Recovery, *ACS Sustain. Chem. Eng.*, 2022, pp. 11191–11202. https://doi.org/10.1021/acssuschemeng.2c02797

[4] P. Thakur, A. Thakur, S. Gautam, J. Choudhary, R. Kumari, K. Raina, R. Sharma and A. Chaudhary, Occurrence and formation of environmentally persistent free radicals in incineration and their impact on soil and water, *J. Geochemical Explor.*, 2023, pp. 107264. https://doi.org.10.1016/j.gexplo.2023.107264

[5] K. A. Omar and R. Sadeghi, Physicochemical properties of deep eutectic solvents: A review, *J. Mol. Liq.*, 2022, pp. 119524. https://doi.org.10.1016/j.molliq.2022.119524

[6] T. El Achkar, H. Greige-Gerges, and S. Fourmentin, Basics and properties of deep eutectic solvents: a review, *Environ. Chem. Lett.*, 2021, pp. 3397–3408. https://doi.org.10.1007/s10311-021-01225-8

[7] G. M. Martínez, G. G. Townley, and R. M. Martínez-Espinosa, Controversy on the toxic nature of deep eutectic solvents and their potential contribution to environmental pollution, *Heliyon*, 2022, pp. e12567. https://doi.org.10.1016/j.heliyon.2022.e12567

[8] E. L. Smith, A. P. Abbott, and K. S. Ryder, Deep Eutectic Solvents (DESs) and Their Applications, *Chem. Rev.*, 2014, pp. 11060–11082. https://doi.org.10.1021/cr300162p

[9] H. Zhang, X. Cui, H. Wang, Y. Wang, Y. Zhao, H. Ma, L. Chai, Y. Wang, X. Hou, and T. Deng, Degradation of polycarbonate-based polyurethane via selective cleavage of carbamate and urea bonds, *Polym. Degrad. Stab.*, 2020, pp. 109342 https://doi.org.10.1016/j.polymdegradstab.2020.109342

[10] O. Długosz and M. Banach, Green methods for obtaining deep eutectic solvents (DES), *J. Clean. Prod.*, 2024, pp. 139914. https://doi.org.10.1016/j.jclepro.2023.139914

[11] R. J. Isaifan and A. Amhamed, Review on Carbon Dioxide Absorption by Choline Chloride/Urea Deep Eutectic Solvents, *Adv. Chem.*, 2018, pp. 1–6. https://doi.org.10.1155/2018/2675659

[12] H. Sui, X. Ju, X. Liu, K. Cheng, Y. Luo, and F. Zhong, Primary thermal degradation effects on the polyurethane film, *Polym. Degrad. Stab.*, 2014, pp. 109–113. https://doi.org.10.1016/j.polymdegradstab.2013.11.021

[13] Y. Wang, H. Song, H. Ge, J. Wang, Y. Wang, S. Jia, T. Deng, and X. Hou, Controllable degradation of polyurethane elastomer via selective cleavage of C-O and C-N bonds, *J. Clean. Prod.*, 2018, pp. 873–879. https://doi.org.10.1016/j.jclepro.2017.12.046

[14] P. Thakur, A. Thakur, S. Gautam, J. Choudhary, R. Kumari, K. Raina, R. Sharma and A. Chaudhary, Occurrence and formation of environmentally persistent free radicals in incineration and their impact on soil and water, *J. Geochemical Explor.*, 2023, pp. 107264. https://doi.org.10.1016/j.gexplo.2023.107264

Frontiers of Chemical and Materials Engineering - ICoFCheM 2025   Materials Research Forum LLC
Materials Research Proceedings 60 (2026) 69-75   https://doi.org/10.21741/9781644903971-10

Structural and Dielectric Analysis of Ba$_{0.6}$Sr$_{0.4}$TiO$_3$ for MLCC Applications

Shinbegar Vashinee Jayasilan[1,2,a], Rozana Aina Maulat Osman[1,2,b*],
Mohd Sobri Idris[2,3,c*], Pagupathi Devandran[1,2,d], Ismail Danish Rozaimia[1,2,e],
Abdul Hadi Kamar[2,3,f], Mohamed Yazid Bin Abdul Latiff[1,2,g],
Thamill Maran Letchmanan[1,2,h], Zurina Shamsudin[2,4,i], and
Yasmin Abdul Wahab[2,5,j]

[1]Faculty of Electronic Engineering Technology, Universiti Malaysia Perlis, 02600 Arau, Perlis, Malaysia

[2]Centre of Excellence for Frontier Materials Research, Universiti Malaysia Perlis, 02600 Arau, Perlis, Malaysia

[3]Faculty of Chemical Engineering Technology, Universiti Malaysia Perlis, 02600 Arau, Perlis, Malaysia. a Faculty of Electronic Engineering Technology, Universiti Malaysia Perlis, 02600 Arau, Perlis, Malaysia

[4]Fakulti Teknologi Kejuruteraan Industri dan Pembuatan, Universiti Teknikal Malaysia Melaka, Hang Tuah Jaya, 76100 Durian Tunggal, Melaka, Malaysia

[5]Nanotechnology & Catalysis Research Centre, University of Malaya, 50603 Kuala Lumpur, Malaysia

[a]jayasilanshin@gmail.com, [b]rozana@unimap.edu.my, [c]sobri@unimap.edu.my,
[d]pagupathi.devandran@gmail.com, [e]ismaildanish1998@gmail.com,
[f]hadikamar@studentmail.unimap.edu.my, [g]yazidlatiff92@gmail.com,
[h]thamill.maran.letchmanan@gmail.com, [i]zurina.shamsudin@utem.edu.my,
[j]yasminaw@um.edu.my

Keywords: Perovskite, BaTiO$_3$, MLCC, X-Ray, Ceramic, Solid State

Abstract. Ba$_{0.6}$Sr$_{0.4}$TiO$_3$ (BST) ceramics were synthesized using the conventional solid-state method and characterized for their dielectric and electrical properties toward multilayer ceramic capacitor (MLCC) applications. X-ray diffraction confirmed a single-phase cubic perovskite structure, indicating successful Sr^{2+} substitution in the Ba^{2+} lattice. The dielectric constant (ε_r) exhibited thermally stable behaviour, decreasing from 2601.69 at 40 $°C$ to 290.72 at 200 $°C$, while dielectric loss ($tan\ \delta$) remained low (0.03-0.05) across the temperature range, reflecting minimal energy dissipation. AC conductivity increased with temperature following Jonscher's power law, suggesting thermally activated charge transport. Impedance analysis revealed a transition from grain-dominated resistivity at low temperature to mixed grain and grain boundary conduction at higher temperatures. The combination of high dielectric stability, low loss, and tunable electrical response demonstrates that Ba$_{0.6}$Sr$_{0.4}$TiO$_3$ is a promising candidate for thermally stable and low-loss dielectric components in advanced MLCC applications.

Introduction

Barium strontium titanate (Ba$_{1-x}$Sr$_x$TiO$_3$, BST) has gained significant attention as a high-k dielectric material due to its excellent tunability, high permittivity, and compatibility with multilayer ceramic capacitor (MLCC) technology [1]. The substitution of Sr^{2+} for Ba^{2+} in BaTiO$_3$ adjusts the Curie temperature and broadens the ferroelectric paraelectric transition region, leading to enhanced thermal stability of dielectric properties [2]. For MLCC applications, dielectric

Frontiers of Chemical and Materials Engineering - ICoFCheM 2025   Materials Research Forum LLC
Materials Research Proceedings 60 (2026) 69-75   https://doi.org/10.21741/9781644903971-10

materials must maintain high capacitance and low dielectric loss across a wide range of frequencies and temperatures to ensure energy efficiency and long term reliability [3].

The dielectric constant and dielectric loss (tan δ) are critical parameters influencing capacitor performance; low dielectric loss reduces energy dissipation and internal heating during high-frequency operation [1],[4]. Several studies have demonstrated that compositional control and rare earth doping in $BaTiO_3$ based ceramics can significantly improve dielectric stability while suppressing loss [5]. The incorporation of Sr^{2+} ions minimizes lattice distortion and enhances the cubic phase stability, resulting in better electrical homogeneity and thermal endurance [6].

Furthermore, microstructural refinement and dopant engineering, such as Mo or Sn modification, have been shown to enhance dielectric constant and reduce defect related conduction in BST ceramics [7],[8]. Gradient composite BST structures also demonstrate excellent permittivity-temperature stability and low dielectric loss, making them suitable for high-frequency MLCC applications [9]. These advancements underline the importance of exploring BST compositions with optimized Sr content to achieve thermally stable, low-loss, and high-k dielectric characteristics for next-generation electronic devices [10].

Experimental Procedure

The $Ba_{0.6}Sr_{0.4}TiO_3$ ceramic was synthesized using the conventional solid-state reaction method, employing high-purity $BaCO_3$, $SrCO_3$, and TiO_2 powders weighed according to the stoichiometric ratio. The powders were thoroughly mixed, and ground using an agate mortar and pestle for 30 minutes to ensure homogeneity and prevent contamination. The resulting mixture was pressed into cylindrical pellets using a uniaxial hydraulic press (2 tonne) and calcined at 1000 °C for 4 hours in air to promote phase formation. After regrinding, the pellets were sintered at 1350 °C for 5 hours to achieve densification. Structural analysis was carried out using X-ray diffraction (XRD) with a Bruker D2 Phaser diffractometer equipped with a LYNXEYE 1D detector and Cu Kα radiation over a 2θ range of 20° to 80°. For electrical measurements, both surfaces of the sintered pellets were coated with silver electrodes and cured at 600 °C for 10 minutes. Dielectric properties such as dielectric constant, dielectric loss, capacitance, and AC conductivity were measured using an LCR meter across a frequency range of 100 Hz to 1 MHz at the temperature of 40°C to 200°C. Impedance spectroscopy was also performed within the same frequency range, and equivalent circuit modelling was conducted using ZView software to analyse the contributions from grains and grain boundaries, confirming the material's potential for multilayer ceramic capacitor (MLCC) applications.

Results

X-Ray Diffraction Analysis

X-ray diffraction (XRD) analysis, as illustrated in Fig. 1, confirms the formation of a single-phase perovskite structure for $Ba_{0.6}Sr_{0.4}TiO_3$, with no detectable secondary phases such as $BaCO_3$, $SrCO_3$, or TiO_2, indicating excellent phase purity. All diffraction peaks are well indexed to the cubic perovskite $BaTiO_3$ phase (PDF No. 00-034-0411), corresponding to the (100), (110), (111), (200), (201), (211), (220), (212), (310), (311), and (222) crystallographic planes. A minor peak splitting observed near $2\theta \approx 45°$ and 65° may indicate slight local strain or partial tetragonal distortion; however, the overall diffraction profile remains consistent with a dominant cubic phase. The substitution of smaller Sr^{2+} ions (1.44 Å) for larger Ba^{2+} ions (1.61 Å) leads to a reduction in lattice distortion, thereby enhancing the structural stability of the cubic phase at room temperature. Furthermore, the sharp and intense diffraction peaks signify a high degree of crystallinity, which is advantageous for achieving stable and reproducible dielectric behaviour in multilayer ceramic capacitor (MLCC) applications.

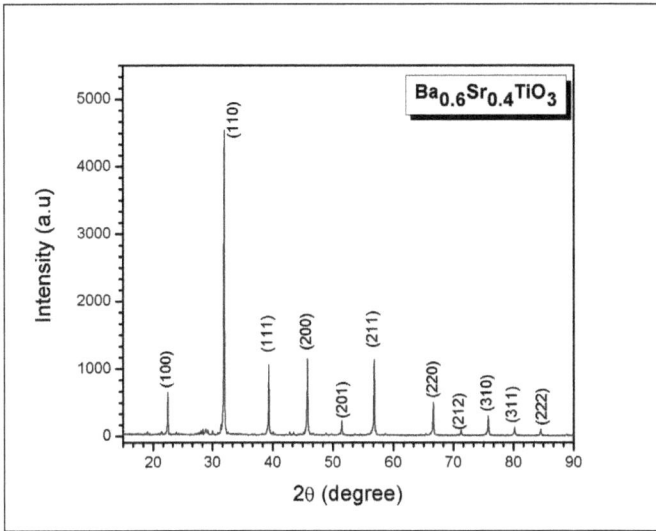

Fig.1. X-ray Diffraction of $Ba_{0.6}Sr_{0.4}TiO_3$

Impedance Spectroscopy Analysis

Fig. 2 presents the frequency dependent capacitance of $Ba_{0.6}Sr_{0.4}TiO_3$ ceramic measured at various heating temperatures ranging from 40 °C to 200 °C. All samples exhibit a typical dielectric dispersion behaviour, where the capacitance decreases progressively with increasing temperature, consistent with thermally influenced dielectric relaxation in perovskite ceramics. At lower temperatures, the sample exhibits the highest capacitance value of approximately 2.04×10^{-8} F at 40 °C, while the lowest capacitance of 2.27×10^{-10} F is recorded at 200 °C. The systematic decline in capacitance with rising temperature suggests a reduction in overall dielectric polarization, likely attributed to diminished dipolar alignment and increased thermal agitation of charge carriers.

Across the measured frequency range (10^2-10^6 Hz), the capacitance values remain nearly constant for all temperatures, indicating minimal frequency dispersion and stable dielectric behaviour. The absence of relaxation peaks or irregular transitions further implies a uniform microstructure and thermally stable dielectric response. Such behaviour confirms that $Ba_{0.6}Sr_{0.4}TiO_3$ possesses excellent temperature and frequency stability, making it a promising candidate for multilayer ceramic capacitor (MLCC) applications, where consistent dielectric performance under varying thermal conditions is essential.

Frontiers of Chemical and Materials Engineering - ICoFCheM 2025   Materials Research Forum LLC
Materials Research Proceedings 60 (2026) 69-75   https://doi.org/10.21741/9781644903971-10

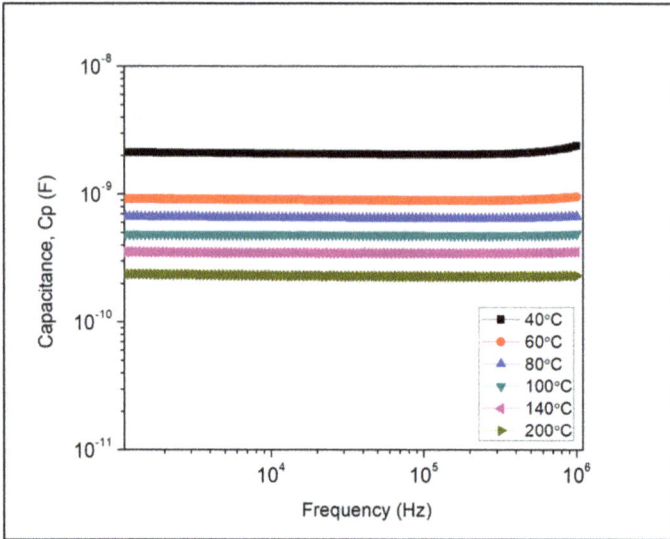

Fig 2. Capacitance vs. Frequency of $Ba_{0.6}Sr_{0.4}TiO_3$ measured from 40°C,
60°C, 80°C, 100°C, 140°C *and* 200°C

Fig. 3 presents the temperature-dependent dielectric constant (ε_r) of $Ba_{0.6}Sr_{0.4}TiO_3$ ceramic measured in the range of 40 °C to 200 °C at multiple frequencies (100 Hz to 1 MHz). The dielectric constant is evaluated using the equation (1).

$$\varepsilon_r = \frac{C.d}{\varepsilon_0.A} \tag{1}$$

where C is the measured capacitance (F), d is the sample thickness (m), A is the electrode area (m²), and ε_0 is the permittivity of free space (8.854×10^{-12} F·m⁻¹).

Fig. 3 presents the temperature dependence of the dielectric constant (ε_r) and dielectric loss (tan δ) of $Ba_{0.6}Sr_{0.4}TiO_3$ measured in the temperature range of 40 °C to 200 °C with the frequency of 100 kHz. The dielectric constant exhibits a gradual decrease with increasing temperature, from approximately $\varepsilon_r \approx 2601.69$ at 40 °C to $\varepsilon_r \approx 290.72$ at 200 °C. Meanwhile, the dielectric loss (tan δ) remains relatively low and stable, with values ranging between 0.03 and 0.05 across the entire temperature range, indicating minimal energy dissipation within the material. The observed decrease in dielectric constant with increasing temperature can be attributed to the temperature induced weakening of dipolar and space-charge polarization mechanisms. At lower temperatures, the alignment of dipoles and charge carriers with the applied electric field is more effective, leading to a higher dielectric response. As temperature increases, thermal agitation disrupts this alignment, thereby reducing overall polarization and consequently lowering the dielectric constant. The relatively low and stable dielectric loss throughout the temperature range signifies low conduction losses and excellent dielectric stability. This behaviour suggests that the ceramic possesses well developed grain boundaries and minimal defect states, which suppress charge carrier migration and reduce leakage currents.

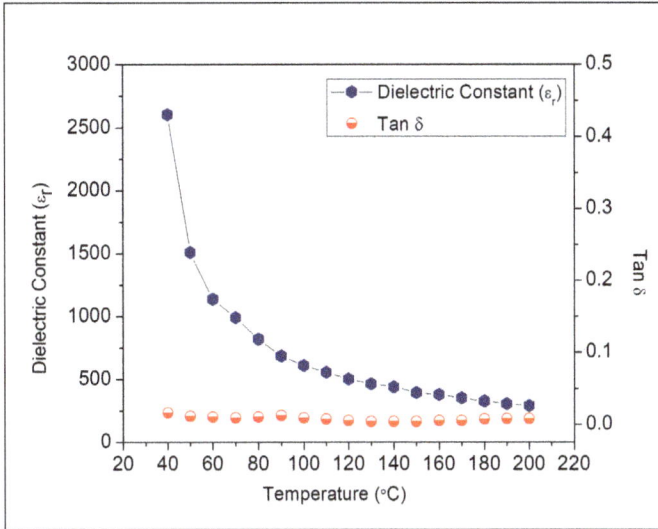

Fig 3. Temperature dependent of the dielectric constant (ε_r) and dielectric loss of $Ba_{0.4}Sr_{0.6}TiO_3$ ceramic

Figure 4 shows the frequency-dependent AC conductivity (σ_{ac}) of $Ba_{0.4}Sr_{0.6}TiO_3$ measured at various temperatures. The plots reveal two distinct region: a low-frequency plateau where σ_{ac} remains nearly constant, corresponding to long range hopping or drift of charge carriers, and a high-frequency dispersion region where σ_{ac} increases sharply with frequency due to short-range hopping of localized charge carriers between defect sites. As temperature increases, the overall conductivity rises, indicating that the conduction mechanism is thermally activated. At 200 °C, the material exhibits the most stable AC conductivity range, with values between 4.08×10^{-9} S·cm^{-1} and 2.63×10^{-6} S·cm^{-1}.

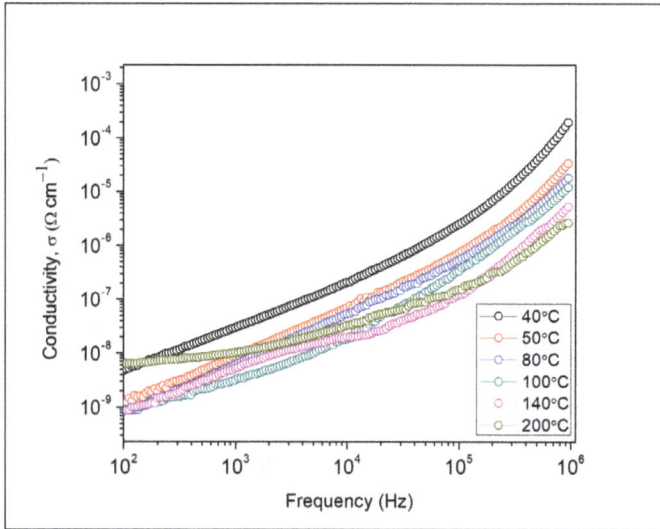

Fig 4. Frequency dependent AC conductivity (σ_{ac}) of $Ba_{0.4}Sr_{0.6}TiO_3$ measured from 40°C, 60°C, 80°C, 100°C, 140°C *and* 200°C

Conclusion

$Ba_{0.4}Sr_{0.6}TiO_3$ ceramics were successfully synthesized, and their dielectric and electrical properties were systematically investigated in the temperature range of 40 °C to 200 °C. The dielectric constant exhibited a smooth, temperature-dependent decrease without any sharp transition, indicating a stable paraelectric phase with weak relaxor like characteristics. This behaviour is associated with the substitution of Sr^{2+} ions at the Ba^{2+} sites, which induces slight lattice distortion and enhances the structural stability of the cubic perovskite phase. The dielectric loss (tan δ) remained low and nearly constant across all measured temperatures and frequencies, confirming the material's low-loss dielectric performance suitable for capacitor applications. The AC conductivity increased gradually with temperature, consistent with a thermally activated hopping mechanism described by Jonscher's universal power law, suggesting enhanced charge carrier mobility at elevated temperatures. Impedance spectroscopy analysis revealed a transition from grain-dominated resistive behaviour at low temperatures to combined grain and grain boundary conduction at higher temperatures, accompanied by a noticeable decrease in total resistance. Modulus spectroscopy indicated a shift from localized to correlated relaxation with increasing temperature, implying improved charge transport pathways and electrical homogeneity within the ceramic.

Overall, the combination of high dielectric stability, low dielectric loss, and thermally activated conductivity demonstrates that $Ba_{0.4}Sr_{0.6}TiO_3$ is a promising candidate for temperature-stable and low-loss dielectric components, particularly in multilayer ceramic capacitor (MLCC) applications.

References

[1] M.R.Z. Nur Farahin Abdul Hamid, R.A.M. Osman, M.S. Idris, "Review on preparation and properties of high-K dielectric material based on Lanthanum doped Barium Titanate," Materials Science Forum, vol. 819, pp. 173-178. https://doi.org/10.4028/www.scientific.net/MSF.819.173

[2] F.A. Ismail, R.A.M. Osman, M.S. Idris, N.A.M. Ahmad Hambali, "Structure and Electrical Characteristics of $BaTiO_3$ and $Ba_{0.99}Er_{0.01}TiO_3$ Ceramics," Solid State Phenomena, vol. 280, pp. 127-133. https://doi.org/10.4028/www.scientific.net/SSP.280.127

[3] M.H. Fisoldin, M.S. Idris, R.A.M. Osman, K.N.D.K. Muhsen, Z.A.Z. Jamal, "Electrical properties of Sn doped $SrTiO_3$," AIP Conference Proceedings, vol. 2203, no. 1, 020044. https://doi.org/10.1063/1.5142136

[4] C. Liu et al., "Improvement of dielectric thermal stability of BST ceramics," Materials Letters, 2011. https://doi.org/10.1016/j.materresbull.2011.04.029

[5] K.G. Kelele et al., "Microstructural and dielectric properties of Mo-doped BST ceramics," Journal of Materials Science: Materials in Electronics, 2022. https://doi.org/10.1080/10667857.2022.2151685

[6] X. Jili et al., "Microstructure and dielectric properties of gradient composite BST MLCCs," Micromachines, vol. 15, no. 4, 2024. https://doi.org/10.3390/mi15040470

[7] S. Pattipaka et al., "Ceramic-based dielectric materials for energy storage applications," Frontiers in Materials, 2024. https://doi.org/10.3390/ma17102277

[8] P. Zhao et al., "Perspectives and challenges for lead-free energy-storage MLCCs," Journal of Advanced Ceramics, 2021. https://doi.org/10.1007/s40145-021-0516-8

[9] X. Huang et al., "Excellent permittivity-temperature stability in BCT/MLCC composites," Journal of Electronic Materials, 2023.

[10] Y. Yang et al., "Giant energy storage density with ultrahigh efficiency in MLCCs," Nature Communications, 2025.

Frontiers of Chemical and Materials Engineering - ICoFCheM 2025　　　　Materials Research Forum LLC
Materials Research Proceedings 60 (2026) 76-82　　　　https://doi.org/10.21741/9781644903971-11

Elucidating the in-situ Integration Mechanisms of Graphene Oxide with BSCF Perovskite for Intermediate-Temperature SOFC Cathodes

M. Darus[1,2,a*], M. Asri Idris[2,3,b], Nur Farhana M. Yunos[1,2,c],
N.A. Mohd Noor[3,d] and A.A. Samat[1,e]

[1]Faculty of Mechanical Engineering & Technology, Universiti Malaysia Perlis (UniMAP), 02600 Arau, Perlis, Malaysia

[2]Frontier Materials Research, Centre of Excellence (FrontMate), Universiti Malaysia Perlis (UniMAP), 02600 Arau, Perlis, Malaysia

[3]Faculty of Chemical Engineering & Technology, Universiti Malaysia Perlis (UniMAP), 02600 Arau, Perlis, Malaysia

[a]murizam@unimap.edu.my, [b]asri@unimap.edu.my, [c]farhanadiyana@unimap.edu.my, [d]noorazira@unimap.edu.my, [e]abdullahabdul@unimap.edu.my

Keywords: IT-SOFC, BSCF, Reduced-Graphene Oxide, X-Ray Photoelectron Spectroscopy (XPS), Binding Energy Shifts

Abstract. This study investigates the in-situ thermal integration of graphene oxide (GO) into $Ba_{0.5}Sr_{0.5}Co_{0.2}Fe_{0.8}O_{3-\delta}$ (BSCF) perovskite cathodes under a nitrogen atmosphere, aiming to enhance intermediate-temperature solid oxide fuel cell (IT-SOFC) performance. Structural and surface analyses using X-ray diffraction (XRD), Raman spectroscopy, and X-ray photoelectron spectroscopy (XPS) reveal that GO undergoes partial reduction to rGO, forming a conductive scaffold that modulates the BSCF matrix. Raman spectra confirmed the evolution of distinct D (~1340 cm^{-1}) and G (~1590 cm^{-1}) bands under reducing conditions, with the highest I_D/I_G ratio observed in N_2-H_2 atmosphere, indicating efficient GO reduction and restoration of sp^2 graphitic domains. In contrast, composites sintered in air showed only perovskite lattice vibrations, confirming suppression of rGO survival. Complementary XPS analysis demonstrated that rGO incorporation promotes oxygen vacancy formation, stabilizes Co^{3+} and Fe^{2+} states, and alters A-site cation environments, as evidenced by binding energy shifts in Ba 3d and Sr 3d spectra. Notably, the Fe^{2+}/Fe^{3+} ratio increased from ~0.69 in pristine BSCF to ~1.04 in the composite, indicating a redox shift facilitated by the modified electronic environment introduced through rGO incorporation. The O 1s spectra showed intensified adsorbed and chemisorbed oxygen peaks (~531.7 eV and ~533.7 eV), confirming enhanced surface reactivity. These modifications collectively suggest improved oxygen ion mobility and potential impact on conductivity. Overall, this work clarifies the integration mechanism of rGO in perovskite oxides and provides mechanistic insights into the design of next-generation IT-SOFC cathodes.

Introduction

The escalating global energy demand and environmental concerns drive the shift towards efficient, sustainable energy systems. Solid Oxide Fuel Cells (SOFCs) are promising due to their high energy conversion efficiency, fuel flexibility, and low emissions [1]. Traditional SOFCs operate at 800 - 1000 °C, which raises material and cost challenges. In contrast, Intermediate-Temperature SOFCs (IT-SOFCs), operating at 500 - 700 °C, reduce material degradation, shorten start-up times, and improve integration with other energy technologies [2].

A key challenge in IT-SOFCs lies in cathode material design, as oxygen reduction reaction (ORR) kinetics and ionic transport become sluggish at lower temperatures. $Ba_{0.5}Sr_{0.5}Co_{0.2}Fe_{0.8}O_{3-\delta}$ (BSCF), a mixed ionic-electronic conducting (MIEC) perovskite, offers high oxygen vacancy

Frontiers of Chemical and Materials Engineering - ICoFCheM 2025   Materials Research Forum LLC
Materials Research Proceedings 60 (2026) 76-82   https://doi.org/10.21741/9781644903971-11

concentration and excellent ORR activity. However, it suffers from surface segregation of Ba/Sr, CO_2 and H_2O sensitivity, and phase instability during thermal/redox cycling, leading to performance degradation [3].

Integrating reduced graphene oxide (rGO) into BSCF has emerged as a strategy to improve conductivity, mechanical stability, and charge transfer [4]. Yet, conventional fabrication methods-such as post-reduction or physical mixing-often yield weak interfacial bonding, agglomeration, and uneven carbon dispersion, especially after high-temperature processing. The functional survival of rGO during sintering under inert or reducing atmospheres, and its role in modulating oxygen vacancies and cation valence states, remain insufficiently understood. Therefore, this study aims to investigate the in-situ thermal reduction of graphene oxide within the BSCF matrix under controlled atmospheres, and to evaluate its effects on phase stability, oxygen vacancy generation, and surface chemical states through XRD, Raman spectroscopy, and XPS analyses. The scope of this work is focused on structural and chemical characterizations, providing mechanistic insights into how rGO integration modifies the BSCF perovskite system, without extending into direct electrochemical performance testing.

Experimental Method

Graphene oxide (GO) was synthesized via a modified Hummers' method, involving the oxidative treatment of graphite flakes in acidic media, followed by purification and freeze-drying. The resulting GO, enriched with oxygen-containing functional groups, was integrated into $Ba_{0.5}Sr_{0.5}Co_{0.2}Fe_{0.8}O_{3-\delta}$ (BSCF) perovskite, which was separately synthesized using a sol-gel combustion method with metal nitrates, citric acid, and ethylene glycol, and subsequently calcined at 950 °C to achieve phase purity and crystallinity.

GO and BSCF powders were homogeneously mixed in a weight ratio of 5:95 (GO:BSCF), ensuring sufficient carbonaceous content for reduction while maintaining perovskite stability. The powders were ball-milled for 12 h in ethanol using zirconia media, dried, and then pressed into pellets under a uniaxial pressure of 200 MPa.

The pellets were sintered at 950 °C for 5 h under a continuous nitrogen flow of 100 mL min^{-1}, with a controlled heating rate of 5 °C min^{-1}. During this high-temperature treatment, GO underwent thermal reduction within the BSCF matrix, leading to the direct formation of rGO inside the composite structure. This process represents the in-situ thermal reduction approach applied in this study, where the transformation of GO to rGO occurs concurrently with the sintering of BSCF. For comparison, additional samples were synthesized under different atmospheres (air and 70% N_2 + 30% H_2), while maintaining identical thermal schedules.

Structural and surface analyses were conducted using X-ray diffraction (XRD) to evaluate phase composition and crystallinity, Raman spectroscopy to assess carbon reduction and perovskite vibrational modes, and X-ray photoelectron spectroscopy (XPS) to analyze oxidation states, oxygen vacancies, and electronic environments of the composite system.

Results and Discussion

X-ray diffraction (XRD) confirmed the successful formation of a single-phase cubic perovskite structure for BSCF (ICDD PDF Card No. 00-055-0563), with sharp reflections indicating high crystallinity. Fig. 1 (a) and (b) shows the SEM-EDS analysis that validated the elemental distribution, showing molar ratios of Ba, Sr, Co, and Fe consistent with theoretical stoichiometry. The X-ray diffraction (XRD) patterns of BSCF-rGO composites sintered under varying partial pressures of oxygen (pO$_2$) are shown in Fig. 1 (c). All samples retained the primary cubic perovskite structure, with no evidence of secondary phases, indicating structural stability upon rGO incorporation. However, subtle shifts in peak positions and variations in full width at half maximum (FWHM), particularly for the (011) reflection, suggest lattice strain and microstructural evolution influenced by the sintering atmosphere. These observations are consistent with the

Frontiers of Chemical and Materials Engineering - ICoFCheM 2025   Materials Research Forum LLC
Materials Research Proceedings 60 (2026) 76-82   https://doi.org/10.21741/9781644903971-11

findings of Sahini et. al., who reported that the unit cell of BSCF expands by approximately 0.8% when annealed at 950 °C under Ar, but contracts by 0.45% under O_2 [5]. The current results similarly demonstrate that pO_2 plays a critical role in modulating the lattice parameters and crystallite coherence of BSCF–rGO composites. The narrower FWHM observed under reducing conditions (e.g., $30H_2/70N_2$) implies enhanced crystallinity or reduced defect density, while broader peaks under oxidizing conditions may reflect increased lattice distortion or oxygen vacancy suppression. These structural responses to pO_2 are crucial for tailoring the electrochemical performance and stability of perovskite-based cathodes.

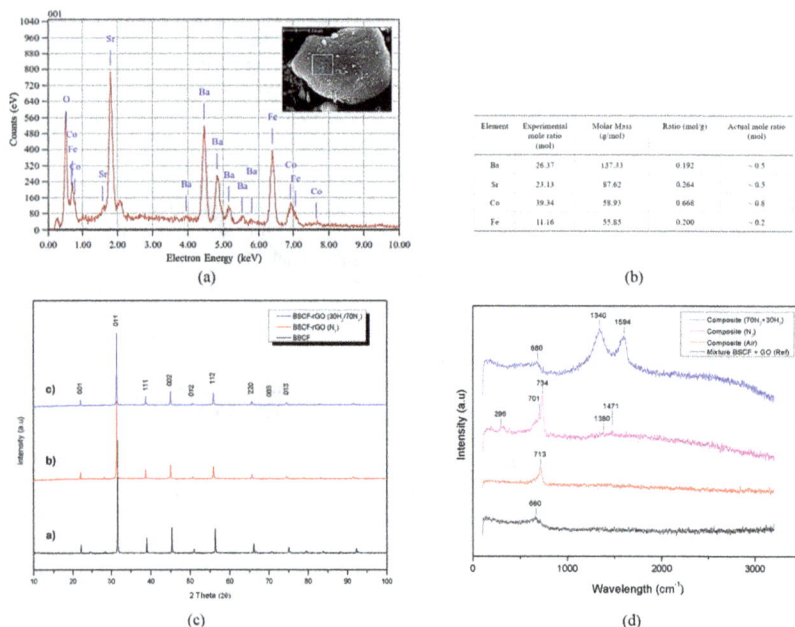

Figure 1 Characterization of BSCF-rGO composites: (a) EDX spectrum with elemental mapping (inset) confirming the presence of Ba, Sr, Co, Fe, and O; (b) elemental composition and calculated molar ratios consistent with BSCF stoichiometry; (c) XRD patterns showing cubic perovskite structure stability under different sintering atmospheres; and (d) Raman spectra highlighting atmosphere-dependent GO reduction, with distinct D and G bands emerging under reducing conditions

Raman spectroscopy provided further insight into the structural evolution of the composites under different sintering atmospheres as shown in Fig 1 (d). The reference mixture of BSCF and GO displayed only a broad band at ~660 cm^{-1}, corresponding to Co/Fe-O stretching vibrations of the perovskite lattice, without the appearance of graphitic D and G bands, confirming that GO was not reduced in the physical mixture. In contrast, composites sintered in air exhibited a strong perovskite vibrational band (~713 cm^{-1}) but no discernible D (~1340 cm^{-1}) and G (~1590 cm^{-1}) bands, suggesting that GO decomposed and the carbon framework did not survive in the oxidizing environment. Under a nitrogen atmosphere, additional features at ~295, 701, and 734 cm^{-1} were observed alongside the emergence of weak D (~1380 cm^{-1}) and G (~1471 cm^{-1}) bands, indicating partial GO reduction and partial retention of sp2 carbon domains. The most distinct reduction effect was recorded in the 70% N_2 + 30% H_2 atmosphere, where pronounced D (~1340 cm[-1]) and G (~1594 cm[-1]) bands were present with an increased I_D/I_G ratio, signifying effective GO reduction and restoration of graphitic domains. These results confirm that rGO incorporation is strongly dependent on the sintering atmosphere, with reducing conditions being essential to preserve the carbonaceous phase. The Raman findings are consistent with XPS analysis, which showed higher sp^2 carbon content and intensified adsorbed oxygen species in composites processed under reducing conditions.

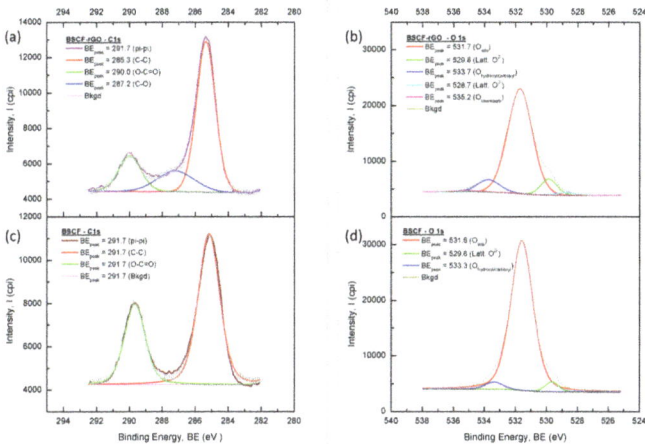

Figure 2 XPS spectral analysis of C1s (a and c) and O1s (b and d) species to compare electronic and surface chemistry properties in BSCF and BSCF-rGO composite.

X-ray photoelectron spectroscopy (XPS) provided further insights into the chemical mechanisms underlying rGO integration within the BSCF matrix. As shown in Fig. 2 (a and c), the C 1s spectra of the BSCF-rGO composite exhibited a pronounced increase in the sp^2 C=C peak (~284.6 eV) accompanied by a notable reduction in oxygenated carbon species (C–O, C=O, O–C=O), indicating effective thermal reduction of GO. This spectral evolution reflects an increase in graphitic carbon content and enhanced electron density on rGO domains. These findings are consistent with previous reports, which demonstrate that the reduction of GO to rGO is typically characterized by a higher proportion of sp^2-hybridized carbon and a corresponding shift in the C 1s spectral profile [6] [7].

Residual oxygen-containing functional groups facilitated hydrogen bonding and electrostatic interactions with Ba, Sr, Co, and Fe cations, thereby enhancing interfacial adhesion and contributing to the stabilization of oxygen vacancies. The O 1s XPS spectra, as shown in Fig. 2 (b and d), exhibited a distinct secondary lattice oxygen peak at approximately 528.7 eV. Additionally, a pronounced adsorbed oxygen peak near 531.7 eV and a chemisorbed oxygen peak around 533.7 eV were observed, which may correspond to surface-bound organic species [8] and hydroxy/carbonyl [9] compounds respectively. These spectral features confirm an increase in surface reactivity and the formation of oxygen vacancies, which are known to promote ionic conductivity and facilitate oxygen ion diffusion [10].

Frontiers of Chemical and Materials Engineering - ICoFCheM 2025   Materials Research Forum LLC
Materials Research Proceedings 60 (2026) 76-82   https://doi.org/10.21741/9781644903971-11

The Ba 3d and Sr 3d spectra as shown in Fig. 3 exhibited subtle shifts in binding energy accompanied by reduced surface concentrations, indicating modifications in the A-site cation environments induced by rGO shielding and electronic redistribution. These changes are likely to enhance thermodynamic stability and facilitate oxygen transport. Preferential formation of oxygen vacancies near the A-site was evident, as the Ba 3d peak shifted toward lower binding energy while the Sr 3d peak moved to a higher binding energy state during composite reduction [11].

The Co 2p spectra as shown in Fig. 4 (a and c) exhibited enhanced Co^{3+} stabilization, while the Fe 2p spectra in Fig. 4 (b and d) revealed a marked increase in Fe^{2+} content, with the Fe^{2+}/Fe^{3+} ratio rising to ~1.04 compared to ~0.69 in pristine BSCF. This increment indicated a redox shift driven by interfacial electron donation from rGO. These spectral changes suggest that rGO plays a dual role in stabilizing Co^{3+} and reducing Fe^{3+}, thereby promoting oxygen vacancy formation and enhancing catalytic activity. The Co 2p energy loss profiles are inherently complex due to multiple electronic transitions, necessitating deconvolution techniques to distinguish intrinsic and extrinsic loss components.

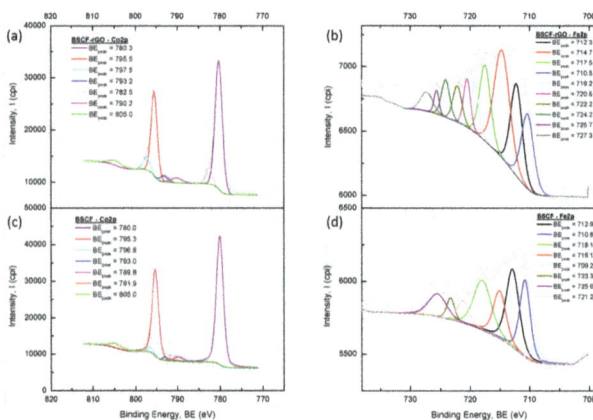

Figure 3 XPS analysis on the deconvoluted Ba3d and Sr3d Spectra occupying the A site of ABO3 perovskite oxide structure to compare electronic and surface chemistry properties in BSCF and BSCF-rGO composite.

Similarly, the Fe 2p spectra are complicated by multiplet splitting and many-electron effects, requiring advanced deconvolution methods for accurate interpretation, as supported by Bagus and Hughes [12] [13]. The presence of multiple peaks identified through automated deconvolution underscores the need for meticulous synthesis and analytical procedures to minimize spectral complexity. Overall, the rGO-induced modifications in Co and Fe chemical states contribute to improved oxygen ion mobility and redox buffering, thereby enhancing the electrochemical performance of the BSCF-rGO composite.

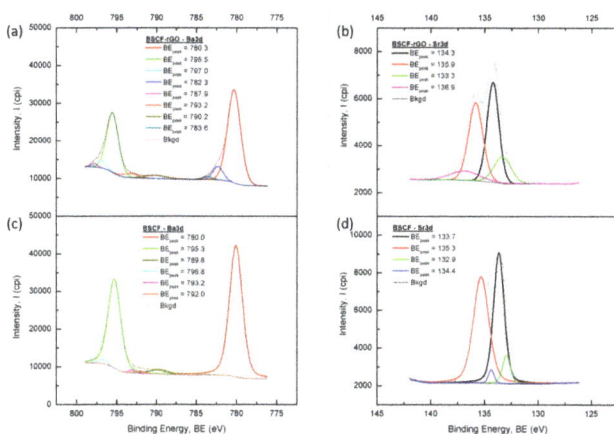

Figure 4 XPS analysis on the deconvoluted Co2p and Fe2p Spectra occupying the B-site of ABO3 perovskite oxide structure to compare electronic and surface chemistry properties in BSCF and BSCF-rGO composite.

Summary

The integration of rGO into the BSCF perovskite matrix via in-situ thermal reduction under inert conditions significantly alters the structural and electronic landscape of the composite. XRD analysis confirmed the retention of the cubic perovskite phase with lattice strain modulated by oxygen partial pressure, while XPS revealed enhanced graphitic carbon content and reduced oxygenated species in the C 1s spectra. The O 1s spectra exhibited distinct peaks at ~528.7 eV, ~531.7 eV, and ~533.7 eV, indicating increased oxygen vacancy formation and surface reactivity. A-site cation analysis showed Ba 3d shifting to lower and Sr 3d to higher binding energies, reflecting preferential oxygen vacancy formation and electronic redistribution. B-site analysis revealed Co^{3+} stabilization and a significant increase in Fe^{2+} content (Fe^{2+}/Fe^{3+} ratio ~1.04), supporting improved redox buffering. These findings demonstrate that rGO serves as both a redox modulator and conductive scaffold, enhancing ionic conductivity and catalytic activity. The mechanistic clarity provided by this study offers a robust framework for optimizing carbon–perovskite interfaces in next-generation IT-SOFC cathodes.

Acknowledgement

The authors gratefully acknowledge the financial support provided by the Fundamental Research Grant Scheme (FRGS) under the Ministry of Higher Education Malaysia (Grant No. FRGS/1/2021/STG05/UNIMAP/02/6) for funding this research project. The authors also extend their sincere appreciation to the Kementerian Pengajian Tinggi (KPT)–Universiti Malaysia Perlis (UniMAP) Skim Latihan Akademik Bumiputera (SLAB) Scholarship for supporting the postgraduate studies that made this work possible.

References

[1] V. Marcantonio and L. Scopel, Thermodynamic Models of Solid Oxide Fuel Cells (SOFCs): A Review, MDPI, Dec. 01, (2024) https://doi.org/10.3390/su162310773

[2] M. F. Vostakola and B. A. Horri, Progress in material development for low-temperature solid oxide fuel cells: A review, Energies (Basel), vol. 14, no. 5, Mar. (2021) https://doi.org/10.3390/en14051280

Frontiers of Chemical and Materials Engineering - ICoFCheM 2025   Materials Research Forum LLC
Materials Research Proceedings 60 (2026) 76-82   https://doi.org/10.21741/9781644903971-11

[3] B. J. Kim *et al.*, Highly Active Nanoperovskite Catalysts for Oxygen Evolution Reaction: Insights into Activity and Stability of Ba0.5Sr0.5Co0.8Fe0.2O2+δ and PrBaCo$_2$O$_{5+δ}$, Adv Funct Mater, vol. 28, no. 45, (2018) https://doi.org/10.1002/adfm.201804355

[4] L. Kashinath and K. Byrappa, Ceria Boosting on In Situ Nitrogen-Doped Graphene Oxide for Efficient Bifunctional ORR/OER Activity, Front Chem vol. 10, (2022) https://doi.org/10.3389/fchem.2022.889579

[5] M. G. Sahini, B. S. Mwankemwa, and N. Kanas, "Ba$_x$Sr$_{1-x}$Co$_y$Fe$_{1-y}$O$_{3-δ}$ (BSCF) mixed ionic-electronic conducting (MIEC) materials for oxygen separation membrane and SOFC applications: Insights into processing, stability, and functional properties, Elsevier Ltd. (2022) https://doi.org/10.1016/j.ceramint.2021.10.189

[6] A. M. Ziatdinov, The Structure and Properties of Graphene Oxide Films and their Changes under High-Temperature Reduction in Inert Atmosphere, (2018) https://www.materialstoday.com/proceedings2214-7853

[7] L. Klemeyer, H. Park, and J. Huang, Geometry-Dependent Thermal Reduction of Graphene Oxide Solid, ACS Mater Lett. (2021) vol. 3, no. 5, pp. 511–515, 2021, https://doi.org/10.1021/acsmaterialslett.0c00423

[8] W. Liu and G. Speranza, Tuning the Oxygen Content of Reduced Graphene Oxide and Effects on Its Properties, ACS Omega (2021) vol. 6, no. 9, pp. 6195–6205 https://doi.org/10.1021/acsomega.0c05578

[9] X. Duan, H. Sun, Z. Ao, L. Zhou, G. Wang, and S. Wang, Unveiling the active sites of graphene-catalyzed peroxymonosulfate activation, Carbon N Y (2016) vol. 107, pp. 371–378 https://doi.org/10.1016/j.carbon.2016.06.016

[10] Y. Yan *et al.*, Oxygen-Vacancy Engineered SnO$_2$ Dots on rGO with N-Doped Carbon Nanofibers Encapsulation for High-Performance Sodium-Ion Batteries, Molecules (2025) vol. 30, no. 15, p. 3203 https://doi.org/10.3390/molecules30153203

[11] C. S. Dandeneau, Y. Yang, B. W. Krueger, M. A. Olmstead, R. K. Bordia, and F. S. Ohuchi, Site occupancy and cation binding states in reduced polycrystalline Sr $_x$Ba$_{1-x}$Nb$_2$O$_6$, Appl Phys Lett. (2014) vol. 104, no. 10 https://doi.org/10.1063/1.4868243

[12] A. E. Hughes, C. D. Easton, T. R. Gengenbach, M. C. Biesinger, and M. Laleh, Interpretation of complex x-ray photoelectron peak shapes. II. Case study of Fe 2p3/2 fitting applied to austenitic stainless steels 316 and 304, Journal of Vacuum Science & Technology A (2024) vol. 42, no. 5 https://doi.org/10.1116/6.0003842

[13] P. S. Bagus, C. J. Nelin, C. R. Brundle, B. V. Crist, N. Lahiri, and K. M. Rosso, Comments on the Theory of Complex XPS Spectra: Extracting Chemical Information from the Fe 3p XPS of Fe Oxides, Taylor and Francis Ltd (2021). https://doi.org/10.1080/02603594.2021.1938007

Frontiers of Chemical and Materials Engineering - ICoFCheM 2025　　　　Materials Research Forum LLC
Materials Research Proceedings 60 (2026) 83-92　　　　https://doi.org/10.21741/9781644903971-12

E-fuels in a Circular Carbon Economy: Insights from Thematic Mapping and Multiple Correspondence Analysis

M.N. Mazlee[1,2,a*] and H. Zunairah[3,b]

[1]Faculty of Mechanical Engineering & Technology, Universiti Malaysia Perlis, 02600 Arau, Perlis, Malaysia

[2]Frontier Materials Research, Centre of Excellence (FrontMate), 01000 Kangar, Perlis, Malaysia

[3]Faculty of Business & Management, Universiti Teknologi MARA, 02600 Arau, Perlis, Malaysia

[a]mazlee@unimap.edu.my, [b]zunairah@uitm.edu.my

Keywords: E-Fuels, Circular Carbon Economy, Hydrogen Based-Synthesis, Thematic Mapping, Multiple Correspondence Analysis

Abstract. Global momentum toward achieving net-zero greenhouse gas emissions has brought electrofuels (e-fuels) to the forefront as practical substitutes for conventional fossil fuels, particularly in sectors where electrification faces major limitations, including aviation, maritime, and heavy industry. Within the Circular Carbon Economy (CCE), e-fuels are generated from renewable hydrogen and captured carbon dioxide (CO_2), offering compatibility with existing fuel systems while enabling large-scale decarbonization. This study employs a bibliometric framework combining annual scientific production, thematic mapping, and multiple correspondence analysis (MCA) to examine the trajectory of e-fuel research from 1984 to 2024. Findings reveal a pronounced increase in publications after 2015, reflecting both rapid technological advances and the influence of international climate policies. Thematic mapping identifies motor themes such as e-fuels, hydrogen, biofuels, and power-to-X, supported by foundational concepts including renewable energy and sustainable economy. Technical niches such as kinetic modelling remain central to process optimization, while emerging themes like NOx reduction highlight evolving environmental concerns. MCA results reinforce a dual research pathway in which one emphasizing technological depth in combustion and electrolysis, and the other focusing on systemic integration through sector coupling and life-cycle assessment. Together, these findings demonstrate that e-fuel research has expanded from fragmented technological explorations into a multidisciplinary field. Positioned at the intersection of engineering, environmental science, and policy, e-fuels are increasingly recognized as pivotal to net-zero strategies, though cost, scalability, and regulatory alignment remain critical challenges for future deployment.

Introduction

Global efforts to achieve net-zero greenhouse gas emissions are reshaping the energy and transportation sectors, accelerating the shift away from fossil fuel dependence. While electrification has delivered meaningful emission reductions across several industries, it remains insufficient for energy-intensive sectors such as aviation, shipping, and heavy-duty transport, where high energy demands and limited storage capacity pose significant barriers [1]. In this context, electrofuels (e-fuels) is the synthetic fuels produced from renewable hydrogen and captured carbon dioxide have emerged as a promising solution. Their ability to integrate with existing infrastructure and contribute to large-scale emission reduction makes them a vital component of the global energy transition [2,3].

Within the CCE framework, e-fuels hold a particularly important role. The CCE model operates on four interconnected principles: reducing, reusing, recycling, and removing carbon [4]. E-fuels embody these principles by combining renewable energy inputs, carbon capture, and

Frontiers of Chemical and Materials Engineering - ICoFCheM 2025   Materials Research Forum LLC
Materials Research Proceedings 60 (2026) 83-92   https://doi.org/10.21741/9781644903971-12

electrochemical conversion processes that transform CO_2 into valuable synthetic fuels. This integration supports circular carbon management while complementing other low-carbon options such as biofuels and hydrogen-based systems [5]. However, achieving widespread adoption depends on continued advances in electrolyser efficiency, declining renewable electricity costs, and stable carbon feedstock supply from both industrial and atmospheric sources [6].

Research on e-fuels has expanded rapidly across disciplines including chemistry, engineering, environmental science, and energy policy. Numerous studies underline their potential to enable deep decarbonization [7], though life-cycle assessments caution that genuine climate benefits are only realized when production is powered entirely by renewable energy [8]. Persistent challenges ranging from production costs to infrastructure readiness continue to shape both academic discourse and policy strategies, particularly in comparison with bioenergy and electrification pathways.

To capture the conceptual and thematic evolution of this expanding field, a bibliometric approach is adopted. This method quantitatively maps research development and interconnections within a domain. In this study, thematic mapping is used to identify established and emerging research themes through keyword co-occurrence analysis, revealing how the field's intellectual structure has matured over time. To complement this, MCA is applied to uncover deeper conceptual relationships, simplifying complex keyword data into visual groupings that illustrate the convergence and diversification of research interests.

By integrating thematic mapping and MCA, this study presents a comprehensive and transparent overview of e-fuel research within the CCE framework. The combined approach traces the intellectual trajectory of the field, highlighting major research foundations, evolving innovation clusters, and critical knowledge gaps that can inform future technological and policy directions in the transition toward a sustainable, low-carbon future.

Methodology

This research employed a quantitative bibliometric approach to trace the conceptual development, thematic evolution, and intellectual structure of e-fuel studies within the context of the CCE. Data were sourced from the Web of Science Core Collection in August 2025 through a Boolean query that included the terms "e-fuels," "electrofuels," "synthetic fuel," and "circular carbon economy." The search was limited to English-language, peer-reviewed journal articles and review papers published between 1984 and 2024. After removing duplicates, irrelevant entries, and non-article records, 399 documents were finalized for analysis.

Bibliographic and citation data were processed using the Bibliometrix R package (version 4.2.3) with support from the Biblioshiny interface. The dataset was meticulously cleaned to ensure accuracy and consistency in author names, institutional affiliations, and keyword formatting. Two categories of keywords were analyzed to represent both deliberate and emerging research trends: author keywords, reflecting the authors' explicit focus, and keywords plus, derived from frequently recurring terms in cited references.

Independent co-occurrence networks were then constructed for each keyword group to visualize thematic connections. These networks were normalized using the association strength technique to correct for frequency disparities and allow meaningful comparison across terms. Thematic mapping was applied to identify dominant research clusters and examine how they evolved and intersected over time, illustrating the intellectual progression of the field.

To further explore conceptual relationships, MCA was conducted on the normalized matrices. MCA projects high-dimensional data into a simplified two-dimensional space, allowing keywords to be positioned according to their co-occurrence behaviour. To enhance reliability and minimize noise, only terms appearing in at least five documents were considered. The resulting coordinates were analyzed using k-means clustering, grouping thematically related keywords based on shared association patterns. The optimal number of clusters was identified through silhouette width

Frontiers of Chemical and Materials Engineering - ICoFCheM 2025   Materials Research Forum LLC
Materials Research Proceedings 60 (2026) 83-92   https://doi.org/10.21741/9781644903971-12

analysis, ensuring both internal cohesion and distinct thematic boundaries. Each cluster was labelled according to its representative keywords and validated through a focused review of relevant literature.

By combining association-based normalization, MCA-driven dimensional reduction, and k-means clustering, this methodological framework ensures a transparent, reproducible, and analytically robust representation of the evolving landscape of e-fuel research within the CCE paradigm.

Results and Discussion

Annual Scientific Production
Over the past four decades, e-fuels research has advanced from a marginal academic topic to one of the most dynamic and rapidly expanding areas in the clean energy field. As illustrated in Figure 1, the number of publications remained minimal for more than twenty years, with only a few papers appearing annually before 2008. A modest rise began in the late 2000s and escalated sharply after 2015, coinciding with intensified global decarbonization efforts and the declining cost of renewable electricity. The most significant growth occurred between 2020 and 2024, when annual publications exceeded one hundred which nearly six times higher than in 2018. This rapid expansion signifies the transition of e-fuels research from a peripheral niche to a central pillar of the global energy transition. It also reflects the increasing interdisciplinary convergence of chemistry, engineering, and policy analysis as researchers collectively address the technical and systemic challenges of scaling e-fuels within the framework of the CCE.

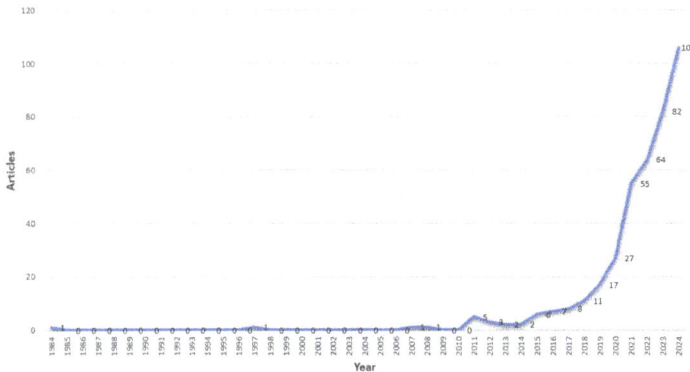

Fig. 1: Annual scientific production of e-fuels publications.

In its early phase, the field concentrated on theoretical and process-level innovations, notably Fischer-Tropsch synthesis and electrolysis integration [6]. As renewable power became more accessible and supportive policies proliferated, research diversified to encompass techno-economic assessments, life-cycle analyses, and system-level evaluations. Reviews by Brynolf et al. [3] and Grahn et al. [5] offered comprehensive evaluations of e-fuel cost and environmental performance across transport sectors, identifying both emerging opportunities and persistent limitations. The surge in publications between 2020 and 2024 corresponds with a broader shift toward cross-sectoral collaboration and integrated mitigation strategies. Oke et al. [1]

demonstrated the potential synergy between electrofuels and biofuels for enhanced emission reduction, while Rennuit-Mortensen et al. [7] analyzed the spatial and energy demands of large-scale deployment. Likewise, Uddin and Wang [8] highlighted the importance of rigorous life-cycle validation, complementing D'Adamo et al.'s [4] review of European policy frameworks supporting e-fuels.

More recent studies have increasingly examined long-term scalability and sustainability concerns. Boretti [2] identified ongoing inefficiencies and infrastructure constraints that continue to challenge commercial deployment. Collectively, these developments reveal that e-fuels research has matured into a robust, interdisciplinary domain situated at the crossroads of technological innovation, policy design, and market transformation. Anchored in the evolving paradigm of the Circular Carbon Economy, it now serves as a critical link between renewable energy innovation and practical decarbonization pathways.

Thematic mapping and Multiple correspondence analysis (MCA)
Combining thematic mapping with MCA provides an integrated perspective on e-fuels research, revealing a solid technological foundation supported by a broad and interconnected thematic landscape.

Thematic mapping
The thematic mapping of authors' keywords (Fig. 2) illustrates how e-fuel research is conceptually organized within the broader framework of the CCE. Research clusters are classified into four categories namely motor, basic, niche, and emerging or declining based on their internal development (density) and overall relevance to the research network (centrality). Both indicators were derived using Callon's method, which quantifies how strongly a theme interacts with others and how cohesive its internal structure is. The resulting two-dimensional strategic diagram visualizes thematic maturity and influence, distinguishing between well-developed, interconnected clusters and those that remain peripheral or evolving.

Motor themes represent the most developed and influential areas, serving as the foundation of e-fuel research. Dominant clusters such as e-fuels, hydrogen, synthetic fuels, methanol, biofuels, and power-to-X form the technological backbone of low-carbon energy innovation. These topics underscore the growing recognition that e-fuels are essential for decarbonizing energy-intensive sectors, including aviation, shipping, and heavy transport [3,5]. Studies by Grahn et al. [5] and Brynolf et al. [3] confirmed their techno-economic feasibility, while Oke et al. [1] highlighted system-wide advantages from combining electrofuels with biofuels.

Basic themes exhibit high centrality but moderate density, signifying their conceptual relevance despite limited technical depth. Clusters such as renewable energy, energy transition, electrification, and sustainable economy serve as the intellectual bridge connecting e-fuel development to broader policy and climate objectives. D'Adamo et al. [4] emphasized that effective governance frameworks, consistent regulation, and economic incentives are key to ensuring that synthetic fuel deployment aligns with long-term decarbonization goals.

Niche themes are technically specialized but less integrated across the research network, reflecting high density and low centrality. These include battery modelling, laminar burning velocity, kinetic modelling, and energy system analysis. Research in this area, such as that of Taherzadeh et al. [6], focuses on improving process efficiency and linking reaction-level modelling with system-level applications through Fischer-Tropsch system optimization.

Emerging or declining themes show both low density and centrality, representing either early-stage explorations or areas of diminishing attention. Topics such as isopropyl alcohol and NOx currently occupy peripheral positions in the research landscape. According to Boretti [2], their

Frontiers of Chemical and Materials Engineering - ICoFCheM 2025   Materials Research Forum LLC
Materials Research Proceedings 60 (2026) 83-92   https://doi.org/10.21741/9781644903971-12

future prominence depends on advances in process efficiency, cost reduction, and environmental performance that may determine whether they evolve into core areas or fade over time.

Overall, the thematic mapping reveals the multidimensional and evolving nature of e-fuel research. The integration of Callon's centrality–density framework provides a clear understanding of how thematic maturity and relevance intersect, highlighting both established knowledge areas and emerging opportunities that continue to shape the trajectory of e-fuel studies within the CCE paradigm.

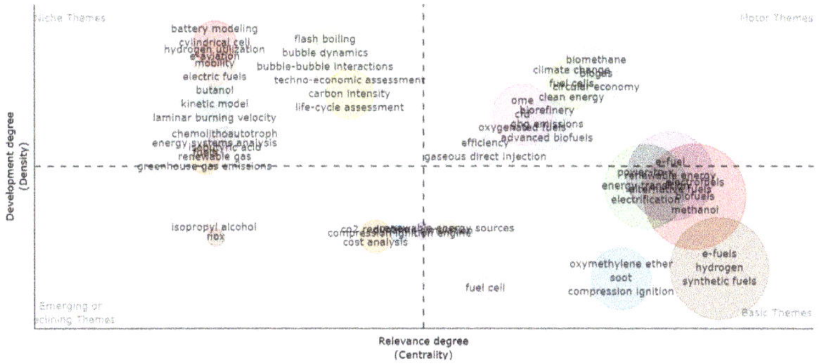

Fig.2: Thematic mapping based authors keywords of e-fuels publications.

The thematic mapping of keywords plus (Fig. 3) provides a comprehensive perspective on how e-fuel research is conceptually structured within the CCE. This analysis complements the study of authors' keywords by uncovering broader methodological and system-level relationships that are often embedded within the literature. The mapping also based on two key parameters: density, which measures a theme's internal development and cohesion, and centrality, which reflects its relevance and connectivity within the wider research network. Both parameters were determined using Callon's method, in which centrality quantifies the degree of interaction between a cluster and others in the network, while density represents the internal strength of relationships among keywords. This centrality-density framework enables a nuanced interpretation of thematic maturity, influence, and evolution. Clusters that are both well developed and strongly interconnected appear as motor themes, whereas those with limited integration or lower development are categorized as emerging or declining.

Frontiers of Chemical and Materials Engineering - ICoFCheM 2025   Materials Research Forum LLC
Materials Research Proceedings 60 (2026) 83-92   https://doi.org/10.21741/9781644903971-12

The spatial distribution of themes in Fig. 3 portrays a research landscape that is technologically sophisticated and rapidly evolving. Motor themes dominate the upper-right quadrant, highlighting clusters such as hydrogen evolution, CO_2 reduction, and H_2 production. These well-established themes underscore hydrogen's dual significance as both a feedstock and an energy carrier in electrofuel synthesis [3,5]. Integrating hydrogen into CO_2 conversion pathways remains fundamental for scaling synthetic fuel production. Rennuit-Mortensen et al. [7] illustrated that large-scale substitution of fossil fuels with electrofuels could reshape renewable energy demand and land-use dynamics, reaffirming hydrogen's cross-sectoral importance within the CCE framework.

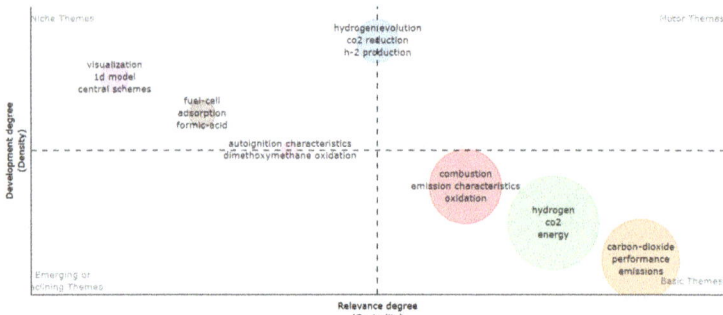

Fig. 3: Thematic mapping based keywords plus of e-fuels publications.

Basic themes occupy a central yet moderately developed position, encompassing clusters such as hydrogen, CO_2, energy, performance, and emissions. These clusters form the conceptual and analytical backbone of e-fuel research, emphasizing techno-economic feasibility, life-cycle assessment, and system-level integration [8]. Studies by Grahn et al. [5] and Oke et al. [1] evaluated the environmental and operational implications of incorporating e-fuels into energy systems, while D'Adamo et al. [4] stressed that coherent policy mechanisms, governance structures, and regulatory stability are essential for long-term sustainability, particularly within Europe.

Niche themes represent highly specialized, method-oriented research domains such as visualization, 1D modelling, and central schemes. Although less connected to the broader research network, these areas contribute to refining simulation tools and improving process optimization. Taherzadeh et al. [6] demonstrated that Fischer-Tropsch system integration supported by advanced simulations can enhance conversion efficiency and reactor performance, linking detailed reaction modelling with system-scale applications.

Emerging or declining themes appear in the lower-left quadrant, characterized by low centrality and density. Topics such as autoignition characteristics and dimethoxymethane oxidation indicate either early-stage explorations or areas losing scholarly momentum. Boretti [2] observed that future advances in efficiency, cost reduction, and environmental performance will determine whether these peripheral themes mature into central research areas or fade from focus.

In summary, the thematic mapping of keywords plus reveals a research field that is consolidating around hydrogen utilization, CO_2 conversion, and system-level performance analysis. The use of Callon's centrality-density framework provides a transparent and structured approach to understanding thematic evolution, highlighting a domain that is both mature and

Frontiers of Chemical and Materials Engineering - ICoFCheM 2025   Materials Research Forum LLC
Materials Research Proceedings 60 (2026) 83-92   https://doi.org/10.21741/9781644903971-12

diversifying which anchored in established technologies while expanding toward new frontiers of innovation within the CCE.

Multiple correspondence analysis (MCA)

The MCA provides a multidimensional understanding of how e-fuel research is conceptually structured within the broader context of the CCE. Applied to both authors' keywords (Fig. 4) and keywords plus (Fig. 5), this approach reveals the underlying conceptual relationships shaping the field's intellectual organization. MCA reduces complex keyword co-occurrence data into a low-dimensional Euclidean space, where each axis represents a principal dimension of conceptual variation. The distance between keywords indicates their frequency of co-occurrence which closer terms imply shared thematic contexts, while greater distances suggest conceptual divergence. K-means clustering was then applied to group terms with similar patterns, allowing clearer identification of major thematic categories. This integrated analysis not only visualizes but also deepens understanding of how technological, environmental, and policy-driven dimensions intersect within the evolving e-fuel research landscape.

In the MCA of authors' keywords (Fig. 4), the two-dimensional map highlights a dual structure linking technological development with system-level integration. Central clusters such as e-fuels, electrofuels, methanol, biofuels, CO_2 emissions, energy efficiency, and techno-economic analysis define the conceptual core of the field. Their close proximity emphasizes the shared focus on cost optimization, energy performance, and environmental evaluation as primary research priorities [3,5]. Brynolf et al. [3] and Grahn et al. [5] recognized that although electrofuels are technologically viable, production costs and environmental uncertainties remain limiting factors. Uddin and Wang [8] further stressed the importance of life-cycle assessments to validate sustainability claims.

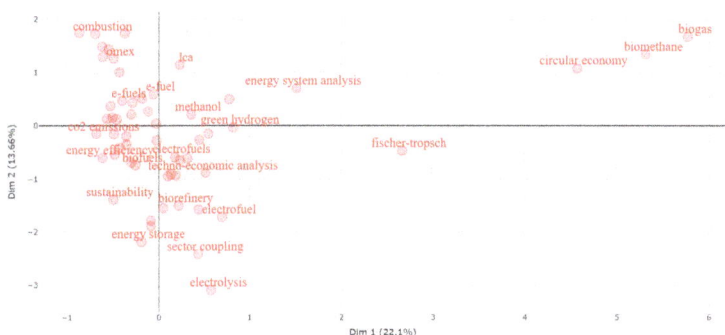

Fig. 4: MCA based authors keywords of e-fuels publications.

On the right-hand side, clusters such as Fischer-Tropsch, biogas, biomethane, and circular economy illustrate how e-fuels are increasingly linked to large-scale deployment and resource circularity [4,6]. These studies highlight the integration of electrofuels into broader decarbonization frameworks. Taherzadeh et al. [6] demonstrated the scalability of Fischer-Tropsch synthesis, while D'Adamo et al. [4] emphasized that coherent policies and governance mechanisms are essential to support adoption in European contexts.

The upper-left quadrant encompasses combustion-related terms such as combustion and Omex, reflecting efforts to balance performance with emissions reduction [2]. The lower-left quadrant,

Frontiers of Chemical and Materials Engineering - ICoFCheM 2025   Materials Research Forum LLC
Materials Research Proceedings 60 (2026) 83-92   https://doi.org/10.21741/9781644903971-12

which includes sustainability, energy storage, and sector coupling, represents system-level concerns related to renewable integration and multi-sector energy systems [1,7].

The MCA of keywords plus (Fig. 5) provides a nuanced interpretation of how e-fuel research is conceptually structured within the CCE. By mapping keyword co-occurrence relationships in a two-dimensional Euclidean space, the analysis highlights how research themes interact and evolve. Dimension 1 (43.9%) captures the continuum from technical process optimization to system-wide integration, while Dimension 2 (13.6%) distinguishes between performance-based experimentation and feedstock development. The proximity of keywords reflects the strength of their conceptual relationship, with clusters identified using the k-means algorithm to distinguish dominant, transitional, and emerging research areas.

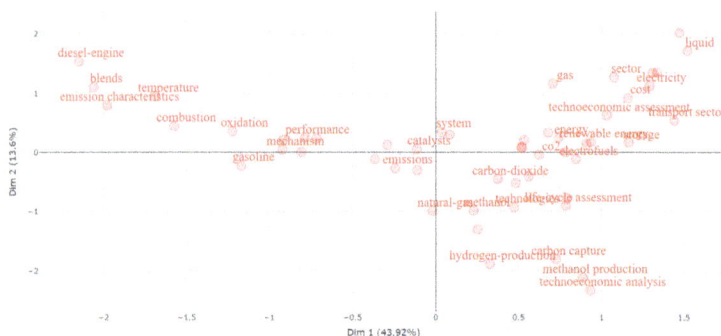

Fig. 5: MCA based keywords plus of e-fuels publications.

The spatial arrangement in Fig. 5 reveals an integrated research landscape that blends performance, cost, and systemic considerations. On the left, clusters such as diesel engine, combustion, oxidation, blends, and emission characteristics represent the experimental foundation of e-fuel studies, emphasizing combustion efficiency and emission control under diverse conditions [2,3]. In the centre, clusters including emissions, catalysts, and carbon dioxide form conceptual bridges linking process-level innovation with environmental evaluation [5,8]. On the right-hand side, clusters such as techno-economic assessment, renewable energy, electricity, transport sector, storage, and cost characterize research aimed at system integration, policy design, and large-scale deployment [1,4]. Finally, the lower-right quadrant, featuring hydrogen production, methanol production, and carbon capture, underscores the feedstock and synthesis aspects vital to scaling Fischer-Tropsch and electrochemical processes [6,7].

Taken together, the MCA results reveal two interrelated research trajectories. One advancing technological refinement through improved combustion, catalysis, and electrolysis, and the other driving system integration through techno-economic analysis, policy coordination, and resource management. Their convergence signals a field shifting from isolated technical investigations toward an interdisciplinary framework that aligns innovation, sustainability, and governance in support of the low-carbon energy transition.

Integrated Discussion
The integration of findings from publication trends, thematic mapping, and MCA provides a cohesive and nuanced understanding of how e-fuel research has evolved within the framework of the CCE. The sharp rise in publications after 2015 marks a pivotal consolidation of the field, driven by advancements in electrochemical conversion, hydrogen production, and Fischer-Tropsch

Frontiers of Chemical and Materials Engineering - ICoFCheM 2025   Materials Research Forum LLC
Materials Research Proceedings 60 (2026) 83-92   https://doi.org/10.21741/9781644903971-12

synthesis. This expansion mirrors global efforts toward achieving net-zero targets and reducing reliance on fossil-based systems [1,5], reaffirming e-fuels as a key solution for decarbonizing hard-to-abate sectors when deployed alongside biofuels for broader, system-level emissions reduction [1].

Thematic mapping identifies e-fuels, hydrogen, methanol, biofuels, and power-to-X as the conceptual anchors of the field [3,5]. These clusters signal a shift from process-specific optimization toward integrative assessments of scalability, cost-effectiveness, and sustainability. Broader themes such as renewable energy, energy transition, and sustainable economy link e-fuel research to global policy frameworks [4], while technical topics like combustion dynamics and kinetic modelling enhance efficiency and performance. The emerging interest in NOx reduction further reflects the ongoing challenge of reconciling energy performance with environmental targets [2,6].

Findings from keywords plus reinforce these observations, emphasizing hydrogen-CO_2 coupling, life-cycle analysis, and system integration as dominant research directions. The prevalence of hydrogen evolution, CO_2 reduction, and hydrogen production confirms hydrogen's central role in synthetic fuel development [3,7]. Clusters on performance, emissions, and policy highlight how technological advancement increasingly depends on governance readiness, financial mechanisms, and cross-sector collaboration [4,8].

MCA results reveal two complementary trajectories driving the field's evolution. The first focuses on technical innovation-combustion enhancement, Fischer–Tropsch optimization, and electrolysis improvement while the second centres on system integration, including techno-economic evaluation, circular economy application, and sector coupling [6,7]. Together, these trajectories show that e-fuel progress depends equally on laboratory innovation and coordinated policy alignment, especially within Europe's CCE initiatives.

By linking bibliometric findings with engineering practice, this analysis demonstrates the field's practical relevance. Advances in combustion and catalytic systems improve efficiency and emissions control, while progress in Fischer-Tropsch and electrolysis technologies enhances production scalability and economic competitiveness. Meanwhile, techno-economic and policy studies ensure that such innovations are embedded within viable infrastructure, regulatory, and market frameworks.

Collectively, the integrated analysis portrays e-fuel research as an evolving, interdisciplinary domain bridging engineering innovation with sustainability governance and industrial implementation. While process optimization remains essential, current research increasingly emphasizes life-cycle validation, policy integration, and system-level adoption. This evolution reflects a mature, strategically coordinated research ecosystem driving the global transition toward sustainable, low-carbon energy systems under the CCE [2,3,5,8].

Conclusions

This bibliometric study, which integrates annual publication trends, thematic mapping, and MCA, provides a comprehensive overview of e-fuels research within the framework of the CCE. The analysis reveals a rapidly expanding field that has progressed from process-level studies to broader, system-wide evaluations linking technology, policy, market dynamics, and sustainability.

Publication output shows an exponential increase after 2015, reflecting the combined influence of international decarbonization commitments and advances in renewable energy technologies. This surge underscores the growing recognition of e-fuels as a critical option for hard-to-abate sectors such as aviation, maritime transport, and heavy industry, particularly when integrated with biofuels to achieve deeper emission reductions.

Thematic mapping identified e-fuels, hydrogen, biofuels, and power-to-X as dominant themes, supported by foundational concepts such as renewable energy and sustainable economy. Technical niches, including Fischer-Tropsch synthesis and kinetic model, remain vital for advancing efficiency, while emerging topics such as NOx reduction reflect evolving environmental priorities and performance challenges.

MCA added further granularity, confirming a dual trajectory in the field. One pathway focuses on technical optimization through combustion, electrolysis, and related processes, while the other emphasizes systemic integration via sector coupling, life-cycle assessment, and circular economy strategies. These findings illustrate that effective deployment of e-fuels requires not only sustained technological innovation but also robust policy support and governance structures.

In sum, e-fuels research is consolidating into a multidisciplinary domain that plays a central role in advancing net-zero strategies. Yet, persistent challenges linked to efficiency, scalability, cost, and regulatory alignment remain. Addressing these issues will be essential for translating academic progress into practical solutions with measurable climate benefits.

References

[1] D. Oke, J.B. Dunn, and T.R. Hawkins, "Reducing economy-wide greenhouse gas emissions with electrofuels and biofuels as the grid decarbonizes," Energy Fuels, vol. 38, no. 7, pp. 6048-6061, 2024. https://doi.org/10.1021/acs.energyfuels.3c04833

[2] A. Boretti, "Reviewing the challenges toward sustainable and carbon-neutral e-fuels," Chem. Ing. Tech., vol. 97, no. 7, pp. 714-725, 2025. https://doi.org/10.1002/cite.202400092

[3] Brynolf, J. Hansson, J. E. Anderson, I. R. Skov, T. J. Wallington, M. Grahn, A. D. Korberg, E. Malmgren, and M. Taljegard, "Review of electrofuel feasibility-prospects for road, ocean, and air transport," Prog. Energy, vol. 4, no. 4, p. 042007, 2022. https://doi.org/10.1088/2516-1083/ac8097

[4] . D'Adamo, M. Gastaldi, and M. Giannini, "Policy frameworks for synthetic fuels in Europe: Opportunities and challenges," Energy Policy, vol. 156, p. 112450, 2021. https://doi.org/10.1016/j.enpol.2021.112450

[5] M. Grahn, E. Malmgren, A. D. Korberg, M. Taljegard, J. E. Anderson, S. Brynolf, J. Hansson, I. R. Skov, and T. J. Wallington, "Review of electrofuel feasibility - cost and environmental impact," Prog. Energy, vol. 4, no. 3, p. 032010, 2022. https://doi.org/10.1088/2516-1083/ac7937

[6] M. Taherzadeh, N. Tahouni, and M.H. Panjeshahi, "Design of an integrated system for electrofuels production through Fischer–Tropsch process," Int. J. Hydrogen Energy, vol. 75, pp. 515–528, 2024. https://doi.org/10.1016/j.ijhydene.2024.03.067

[7] A.W. Rennuit-Mortensen, K.D. Rasmussen, and M. Grahn, "How replacing fossil fuels with electrofuels could influence the demand for renewable energy and land area," Smart Energy, vol. 10, p. 100107, 2023. https://doi.org/10.1016/j.segy.2023.100107

[8] M.N. Uddin and F. Wang, "Fuelling a clean future: A systematic review of techno-economic and life cycle assessments in e-fuel development," Appl. Sci., vol. 14, no. 16, p. 7321, 2024. https://doi.org/10.3390/app14167321

Frontiers of Chemical and Materials Engineering - ICoFCheM 2025 Materials Research Forum LLC
Materials Research Proceedings 60 (2026) 93-100 https://doi.org/10.21741/9781644903971-13

Characterization of Biodegradable Magnesium Tin Oxide as a Preliminary Study for 3-D Printing of Bio-Based Filament Feedstock

R.A. Malek[1,2,a] *, N.E.R. Azman[1,b], S.H.M. Salleh[1,3,c] and
Nur Hidayah Ahmad Zaidi[1,2,d]

[1]Faculty of Chemical Engineering & Technology, Universiti Malaysia Perlis (UniMAP), Taman Muhibbah, 02600 Jejawi, Perlis, MALAYSIA

[2]Center of Excellence Frontier Materials Research, 01000 Seriab, Perlis, MALAYSIA

[3]Center of Excellence Geopolymer and Green Technology, Taman Muhibbah, 02600 Jejawi, Perlis, MALAYSIA

[a]rohayamalek@unimap.edu.my, [b]ellissaazman@gmail.com, [c]sitihawa@unimap.edu.my, [d]hidayah@unimap.edu.my

Keywords: Magnesium Tin Oxide, Biodegradable Filament, 3D Printing Biofilament, Biomaterials Application

Abstract. Recent research has introduced a new class of implant materials made from degradable metals, offering superior initial stability for bone implants. The unique combination of mechanical, electrochemical, and biological properties in these Mg-based Metal Matrix Composites (MMCs) makes them an increasingly popular choice for biodegradable biomaterials. This study focused on developing biodegradable Magnesium Tin Oxide ($MgSnO_3$) as a preliminary step toward fabricating Three-Dimensional (3D) printing biofilament feedstock. Accordingly, samples were fabricated using powder metallurgy techniques, with Sn additions ranging from 3 to 5 wt.%, compacted at approximately 1,500psi, and sintered at 300°C with a heating rate of 10°C/min. Pure Magnesium (Mg), which contains 100% Mg powder, served as the control sample. The physicochemical properties of the samples were analyzed through microstructural observation, thermal decomposition behavior, and immersion tests in simulated body fluid in a controlled environment. Overall, these analyses provided insight into the microstructure of materials, thermal stability, and biocompatibility, which are crucial for their potential use in medical implants. The findings reveal that the addition of wt.% Tin (Sn) to Mg powder affected the physical properties by inducing phase segregation, stabilizing the material, reducing thermal degradation, and preventing cracking during immersion. However, adding 4 wt.% Sn and above led to contradictory results, with negative impacts on the material's performance.

1. Introduction

In the early twentieth century, Magnesium (Mg) alloys and composites were introduced as degradable metal implants in orthopedic and trauma surgery [1]. Due to their intriguing combination of mechanical, electrochemical, and biological properties, Mg composites and their compounds have sparked growing interest as potential biomaterials in the applications of biodegradable implants. This includes stents in vascular medicine or fixation devices for fractured bones in osteosynthesis [1]-[3]. Accordingly, Mg dissolves spontaneously in bodily fluids and has mechanical characteristics comparable to real bone [1].

The increasing demand for sustainable and eco-friendly materials in modern industries has encouraged significant developments in Three-Dimensional (3D) printing technology, particularly in the field of bio-based filament feedstocks. Traditional filaments, often derived from petroleum-based plastics, pose substantial environmental concerns resulting from their non-biodegradable nature and reliance on fossil fuels. Due to advancements in the 3D printing industry, there is an

Frontiers of Chemical and Materials Engineering - ICoFCheM 2025
Materials Research Proceedings 60 (2026) 93-100

Materials Research Forum LLC
https://doi.org/10.21741/9781644903971-13

urgent demand to develop alternative materials that offer sustainable and enhanced performance characteristics. Magnesium Tin Oxide (MgSnO₃) has garnered interest as a promising biodegradable and eco-friendly material due to its exceptional properties, such as high thermal stability, biocompatibility, and potential for biodegradability [3], [4]. Its incorporation into 3D printing feedstocks offers opportunities for developing filaments that provide both environmental and functional benefits [5]. However, its suitability as a bio-based filament feedstock for 3D printing requires further investigation into its physical, chemical, and mechanical properties.

This preliminary study aims to characterize biodegradable $MgSnO_3$ as a potential feedstock for 3D printing, focusing on its processing, structural integrity, and biodegradability. The findings will provide a framework for future research into optimizing a foundation for $MgSnO_3$-based filaments. This, ultimately, contributes to the development of sustainable 3D printing materials with broad industrial applications.

2. Materials and methodology

The preliminary stage of the $MgSnO_3$ Metal Matrix Composite (MMC) involves a structured approach to identifying the optimal mixing composition for biofilament feedstock production. Concurrently, this study is divided into two stages: the fabrication of samples and the testing of these samples to evaluate their physicochemical properties.

2.1 Raw materials

The raw materials used in this study include Mg powder, which serves as the base material for the $MgSnO_3$ MMC, and Tin (Sn) powder. These compounding elements are incorporated to improve the mechanical properties, thermal stability, and degradation behavior of the tested samples. At the same time, a microstructural analysis was conducted using a Scanning Electron Microscope with Energy Dispersive Spectroscopy (SEM-EDS) and the JEOL JSM-6460 LA model. The Mg powder, appearing dark grey, exhibited irregular, angular shapes with a slightly rough surface with a varied particle size distribution, averaging 405.31μm. In contrast, the Sn powder displayed in light grey has near-spherical or spherical shapes with smooth surface textures with approximately 157.29μm of the particle size.

2.2 Fabrication of tested samples using the powder metallurgy technique

Note that all samples were fabricated using the powder metallurgy method, which consists of mixing, compacting, and sintering processes. As presented in Table 1, Sn powder was added between 3 and 5wt.% to observe its behavior. For compacting, a hydraulic hand press machine model VT MHP-10T was used, and all the mixtures were compacted at $1500 \pm 50psi$. Following this, the green compact samples with a diameter of 12mm were processed for sintering at 300°C and at a heating rate of 10°C/min under a controlled environment using an inert gas.

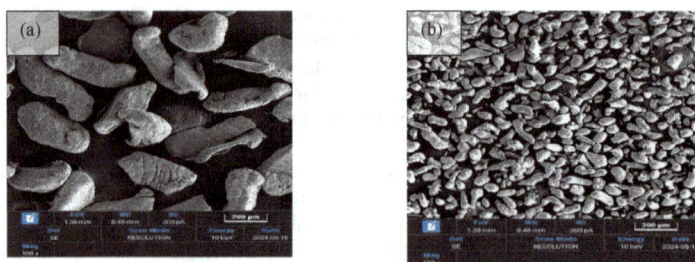

Figure 1. SEM micrograph of (a) Mg and (b) Sn powder at 100X magnification.

Frontiers of Chemical and Materials Engineering - ICoFCheM 2025   Materials Research Forum LLC
Materials Research Proceedings 60 (2026) 93-100   https://doi.org/10.21741/9781644903971-13

Table 1. The mixing design for the samples.

Sample ID	Mg (wt.%)	Sn (%)
Control sample	100	0
Mg with 3wt.% Sn	97	3
Mg with 4wt.% Sn	96	4
Mg with 5wt.% Sn	95	5

2.3 Thermal degradation behavior testing

A sample of pulverized powder of all samples, weighing approximately 5 g, was placed in a non-reactive pan prior to being placed in the Thermogravimetric Analyzer (TGA) instrument Model TGA/DSC1 produced by Mettler Toledo instrument. The temperature was set between 30 and 350°C under a nitrogen environment with a heating rate of 10°C/min. The system began heating the sample at the programmed rate while monitoring both weight changes.

2.4 Immersion testing under simulated body fluid

The approach outlined involves immersing the samples in a supersaturated phosphate solution, specifically Phosphate Buffered Saline (PBS), to mimic the early stages of bone development in the human body. In order to prepare the solution, five PBS tablets were dissolved in one liter of distilled water using a stirring machine at room temperature to ensure a homogeneous mixture. The solution was then placed in a digital ultrasonic bath, maintained at a temperature of 37°C and a pH of 7.4, conditions critical for simulating the normal physiological environment. Subsequently, the samples were incubated in the SBF solution for 7 days within a water bath.

3. Results and discussion

Figures 2(a)-(d) display the SEM images of all tested samples at a magnification of 100X, revealing distinct features corresponding to the sintering temperature used. The microstructures show two distinct regions: a dark region representing the Mg content and a light region indicating the Sn content, which highlights phase segregation between Mg and Sn. This phase segregation provides insight into the elemental distribution across the samples. This phase segregation in alloys has been affected the material strength and ductility as these phases can lead to poor bonding between the elements resulted in lower overall performance of the material under stress. The presence of these segregated phases might affect the material's strength and ductility, as the Mg and Sn phases may not interact effectively to create a homogeneous bond, which could reduce the material's mechanical performance in certain applications. In addition, this phase segregation provides valuable insight into the elemental distribution across the samples, which could influence the diffusion of ions in bioimplant applications or the overall strength of the material under stress. Overall, all the samples presented similar microstructures, suggesting consistent sintering conditions. Additionally, Magnesium Oxide (MgO) formation was identified on the surfaces of the samples, resulting from the high reactivity of Mg, which is susceptible to oxidation under the experimental conditions [6], [7]. This oxide layer is notable as it can influence the mechanical properties of the material, as well as its thermal and chemical stability. The MgO layer could affect the thermal and chemical stability of the material, potentially limiting its effectiveness in high-temperature or harsh environments, as well as reducing the bonding between the Mg and Sn phases, exacerbating phase segregation and affecting the overall material performance. In terms of Sn content, the addition of 3 and 4 wt.% Sn did not result in visible cracking, and the microstructures remained unaffected. However, as illustrated in Figure 3, increasing the Sn content to 5 wt.% caused significant cracking, indicating a shift in material behavior. In other words, this higher Sn content likely modified the microstructure, potentially making the material more brittle and inducing internal stresses, which can lead to cracking. Additionally, the increased Sn content might disrupt the bonding between the Mg and Sn phases, exaggerating the formation of cracks under sintering conditions.

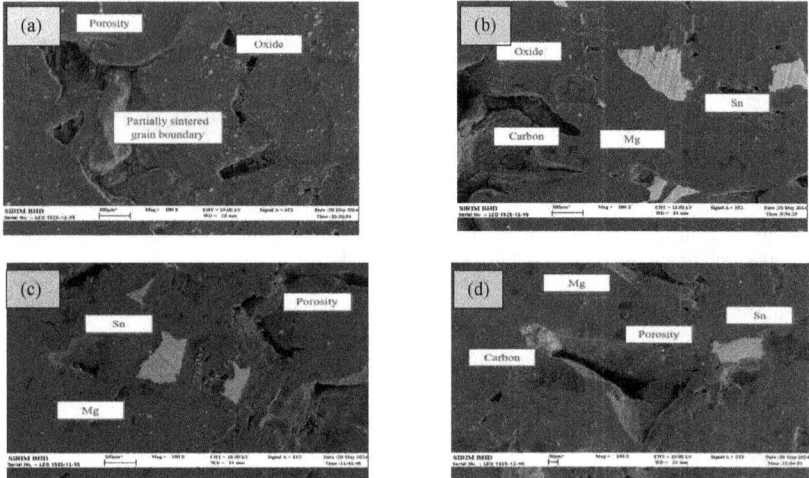

Figure 2. The SEM images of (a) control, (b) Mg with 3 wt.% Sn, (c) Mg with 4 wt.% Sn, and (d) Mg with 5 wt.% Sn samples at 100X magnification showing distinct microstructural features that correspond to the sintering temperature used.

Figure 3. Observation of the sign of cracking for Mg with 5 wt.% Sn sample at 500X magnification.

The dimensional and weight changes observed in the Mg-Sn composite samples offer valuable insights into the effects of Sn addition on their physical properties after sintering. As summarized in Table 2, the control sample exhibited the largest thickness change, with similar trends observed in weight changes. The control sample demonstrated a weight change of 0.37%, while the addition of 3 wt.% Sn reduced this to 0.16%, suggesting that Sn stabilizes the material and reduces volatilization compared to Mg. This effect is likely due to Sn's ability to reduce the oxidation of Mg, helping maintain a more stable mass during sintering [6]. Meanwhile, for the 4 wt.% Sn sample, the weight change was nearly identical to the control, indicating that while Sn addition provides some stabilization, it does not significantly hinder volatilization. Conversely, the sample with more than 5 wt.% Sn exhibited the highest dimensional and weight changes, likely due to internal stresses caused by phase formation, which contributed to mass loss during sintering.

Frontiers of Chemical and Materials Engineering - ICoFCheM 2025   Materials Research Forum LLC
Materials Research Proceedings 60 (2026) 93-100   https://doi.org/10.21741/9781644903971-13

Table 2. Data on dimensional and weight changes for all samples.

Sample ID	Thickness before sintering (mm)	Thickness after sintering (mm)	Thickness change (%)	Weight before sintering (g)	Weight after sintering (g)	Weight change (%)
Control sample	6.80	6.20	8.82	1.1060	1.1019	0.37
Mg with 3 wt.% Sn	6.40	6.20	3.13	1.0891	1.0874	0.16
Mg with 4 wt.% Sn	6.10	5.90	3.27	1.0926	1.0885	0.36
Mg with 5 wt.% Sn	6.10	5.90	3.27	1.0931	1.0880	0.46

The TGA results demonstrate how Sn content influences weight loss during heating. The control sample (0 wt.% Sn) recorded a 2.2% weight loss. Meanwhile, the addition of 3 wt.% Sn reduced the weight loss to 0.87%, indicating that this alloy remains thermally stable with minimal decomposition. The thermal stability in Mg alloys is important in 3D printing where high temperatures are encountered during printing processes. However, when the Sn content was increased to 4 wt.%, the weight loss increased to 2.51%, similar to that of the control sample, suggesting that a higher Sn content leads to greater thermal instability. Despite higher mass loss in the higher Sn samples, the final residue remained similar across all samples, with the residue for the 5 wt.% Sn sample (4.66mg) closely matches that of the 3 wt.% Sn sample (4.68mg).

In the Differential Scanning Calorimetry (DSC) analysis, the phase transition temperature decreased as Sn content increased. The 3 wt.% Sn alloy exhibited the highest transition temperature at 214.49°C, while the 4wt.% Sn alloy's transition temperature dropped significantly to 61.6°C. Notably, the 5 wt.% Sn alloy had a transition peak at 135.3°C, higher than the 4 wt.% sample but still much lower than the 3 wt.% sample. Additionally, the 5wt.% Sn sample absorbed more energy during the phase transition, with an integral normalized value of 287.66J·g⁻¹, indicating a more intense or complete phase change. Collectively, these results suggest that while the melting temperature decreases with increasing Sn content, the transition becomes more significant in terms of energy absorbed. This findings are consistent with recent studies by Marimuthu et.al. (2025) [8], which report similar trends in phase transition temperatures and energy absorption in magnesium alloys with varying Sn content

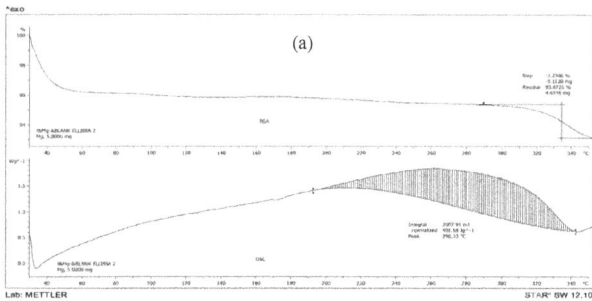

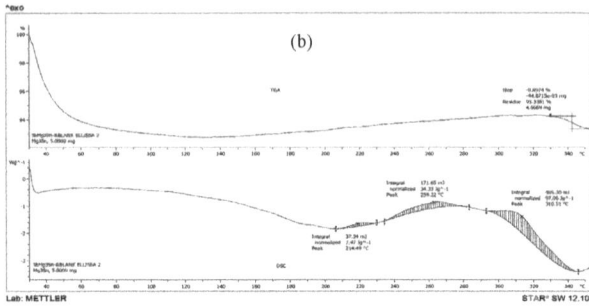

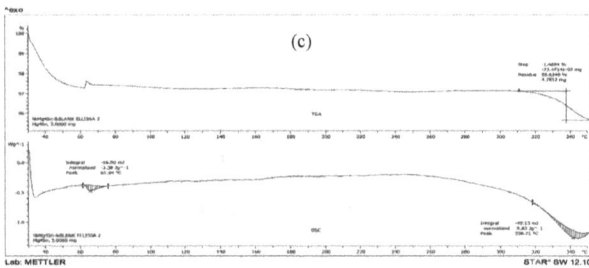

Figure 4. The thermal behavior of the samples tested between 30 and 350°C for (a) control, (b) Mg with 3 wt.% Sn, (c) Mg with 4 wt.% Sn, and (d) Mg with 5 wt.% Sn due to the effect of Sn content.

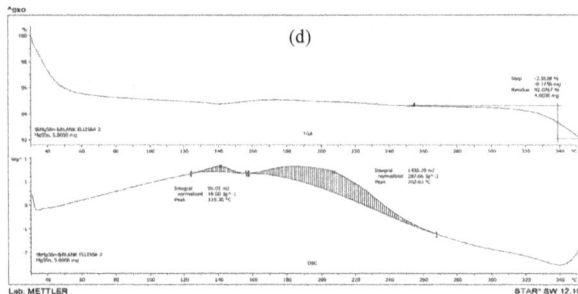

Figure 4. (cont.) The thermal behavior of the samples tested between 30 and 350°C for (a) control, (b) Mg with 3 wt.% Sn, (c) Mg with 4 wt.% Sn, and (d) Mg with 5 wt.% Sn due to the effect of Sn content.

After being immersed in PBS solution for seven days, the samples exhibited significant changes in appearance and weight gain. Among the assessed samples, the control sample yielded the highest weight gain of 30.51%, followed by Mg with 5 wt.% Sn (23.22%), Mg with 4 wt.% Sn (21.63%), and Mg with 3 wt.% Sn (20.04%). As displayed in Figure 5, most samples demonstrated cracking, and some partially dissolved in the PBS solution. All samples underwent corrosion and degradation, with calcium phosphate forming a heavy coating on their outer layer [9]. Additionally, the control sample dissolved the most, while samples with more than 4 wt.% Sn

98

exhibited cracks in their sintered bodies. Surprisingly, the sample with 3 wt.% Sn exhibited no signs of cracking during the immersion period.

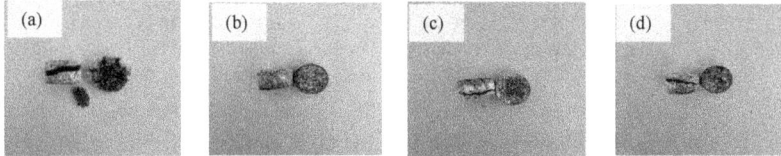

Figure 5. Image of degradation condition during immersion test for (a) control, (b) Mg with 3 wt.% Sn, (c) Mg with 4 wt.% Sn, and (d) Mg with 5 wt.% Sn.

4. Conclusion

This study investigates the effects of Sn addition to Mg-based composites and its influence on the material properties during sintering and degradation in PBS solution. The results reveal that the addition of Sn improved the thermal stability and reduces the volatilization of Mg, with 3 wt.% Sn demonstrates the most promising stability in terms of dimensional and weight changes. The addition of Sn at this stage helps to maintain the intergrity of material particularly at high-tempeature environments. However, higher Sn contents over 4 wt.% led to increased thermal instability, cracking, and phase segregation, suggesting a more brittle microstructure. Additionally, the corrosion and degradation behavior in PBS solution highlighted the significant impact of Sn content on material integrity, with samples containing more than 4 wt.% Sn exhibiting cracks. Thus, these findings suggest that while Sn can enhance the stability of Mg-based composites, careful control of Sn content was crucial to balance mechanical integrity and biodegradability for potential use in medical applications such as biodegradable implants.

Acknowledgement

We would like to extend our appreciation for the support from the Tin Industry (Research and Development) Board Under Grant Number 9025-0019, from the Kementerian Sumber Asli, Alam Sekitar Dan Perubahan Iklim.

References

[1] Pascual-González, Cristina & Thompson, Cillian & Vega, Jimena & Biurrun Churruca, Nicolás & Fernández Blázquez, Juan & Lizarralde, Iker & Herráez- Molinero, Diego & González, Carlos & LLorca, Javier. (2021). Processing and properties of PLA/Mg filaments for 3D printing of scaffolds for biomedical applications. Rapid Prototyping Journal. ahead-of-print. https://doi.org/10.1108/RPJ-06- 2021-0152

[2] Orellana-Barrasa, Jaime, Ana Ferrández-Montero, Aldo. R. Boccaccini, Begoña Ferrari,andJoséYgnacio Pastor(2022)TheMechanical,Thermal,andChemical Properties of PLA-Mg Filaments Produced via a Colloidal Route for Fused- Filament Fabrication. Polymers 14, no. 24: 5414. https://doi.org/10.3390/polym14245414

[3] Hou, L., Li, Z., Zhao, H., Pan, Y., Pavlinich, S., Liu, X., ... Li, L. (2016). Microstructure, Mechanical Properties, Corrosion Behavior and Biocompatibility of As-Extruded Biodegradable Mg–3Sn–1Zn–0.5Mn Alloy. Journal of Materials Science & Technology, 32(9), 874–882. https://doi.org/10.1016/j.jmst.2016.07.004)

[4] Yi-hao Luo, Wei-li Cheng, Hang Li, Hui Yu, Hong-xia Wang, Xiao-feng Niu, Li- fei Wang, Zhi-yong You, Hua Hou. Achieving high strength-ductility synergy in a novel Mg–Bi–Sn–Mn

alloy with bimodal microstructure by hot extrusion, Materials Science and Engineering: A, Volume 834, 2022, 142623, ISSN 0921- 5093, https://doi.org/10.1016/j.msea.2022.142623

[5] Xiaoru Zhuo, Liyan Zhao, Wei Gao, Yuna Wu, Huan Liu, Peng Zhang, Zhichao Hu, Jinghua Jiang, Aibin Ma, Recent progress of Mg–Sn based alloys: the relationship between aging response and mechanical performance, Journal of Materials Research and Technology, Volume 21, 2022, Pages 186-211, 2238-7854, https://doi.org/10.1016/j.jmrt.2022.08.126

[6] Ali F, Al Rashid A, Kalva SN, Koç M. Mg-Doped PLA Composite as a Potential Material for Tissue Engineering—Synthesis, Characterization, and Additive Manufacturing. Materials. 2023; 16(19):6506. https://doi.org/10.3390/ma16196506

[7] Dev, A., Naskar, N., Kumar, N., Jena, A., & Paliwal, M. (2019). A systematic investigation of secondary phase dissolution in Mg–Sn alloys. Journal of Magnesium and Alloys, 7(4), 725–737. https://doi.org/10.1016/j.jma.2019.11.002

[8] R. M., S. Marimuthu, M. Natarajan, S. H V, P. B a, and M. I. Ammarullah, "Effects of Si, Sn, Sr, Zn, and Zr on microstructure and properties of magnesium alloys for biomedical applications: a review," *Philos Mag Lett*, vol. 105, Jul. 2025, https://doi.org/10.1080/09500839.2025.2525095

[9] Soleymani, F., Emadi, R., Sadeghzade, S., & Tavangarian, F. (2020). Bioactivity Behavior Evaluation of PCL-Chitosan-Nanobaghdadite Coating on AZ91 Magnesium Alloy in Simulated Body Fluid. Coatings, 10(3), 231. https://doi.org/10.3390/coatings10030231

Frontiers of Chemical and Materials Engineering - ICoFCheM 2025   Materials Research Forum LLC
Materials Research Proceedings 60 (2026) 101-107   https://doi.org/10.21741/9781644903971-14

Systematic Literature Review on AZ91-CNT Magnesium Nanocomposites: Microstructure, Mechanical and Thermal Performance

Husna Mat Salleh[1,a], Nur Hidayah Ahmad Zaidi[1,b*], Nur Maizatulshima Adzali[1,c], Siti Hasanah Osman[2,d], Sinar Arzuria Adnan[1,e] and Ahmad Mujahid Ahmad Zaidi[3,f]

[1]Faculty of Chemical Engineering & Technology, Universiti Malaysia Perlis, Malaysia

[2]Fuel Cell Institute, Universiti Kebangsaan Malaysia, Malaysia

[3]Universiti Pertahanan Nasional Malaysia, Malaysia

[a]husna6883@gmail.com,[b]hidayah@unimap.edu.my,[c]shima@unimap.edu.my, [d]hasanah@ukm.edu.my, [e]sinar@unimap.edu.my, [f]mujahid@upnm.edu.my

Keywords: AZ91 Magnesium Alloy, Carbon Nanotubes (CNTs), Powder Metallurgy, Mechanical Properties, Thermal Conductivity

Abstract. This paper reviews recent progress on enhancing AZ91 magnesium alloys through carbon nanotube (CNT) reinforcement to meet industrial demands for lightweight, high-strength materials. Following the PRISMA 2020 guidelines, twenty-three peer-reviewed studies (2010–2025) were analyzed to evaluate fabrication methods, microstructural evolution, and resulting property improvements. Powder metallurgy and spark plasma sintering produced the most significant enhancements, achieving up to 36 % higher tensile strength and 86 % greater elongation. Surface coatings such as Ni, MgO, and Pt improved CNT dispersion and interfacial bonding.. Enhanced thermal conductivity and corrosion resistance further broaden application potential in aerospace and automotive sectors. However, limited standardization, scalability issues, and insufficient long-term data remain major challenges. Future research should focus on optimizing processing parameters, hybrid reinforcements, and cost-effective manufacturing to advance AZ91–CNT composite technology.

Introduction

Magnesium (Mg) alloys have gained increasing attention as lightweight structural materials, particularly in the automotive and aerospace industries, where weight reduction without compromising strength is essential [1]. Among these alloys, AZ91 has been widely adopted due to its excellent cast ability and cost-effectiveness [2]. However, despite these advantages, AZ91 suffers from low ductility, poor wear resistance, and limited thermal stability, restricting its use in high-performance applications [3]. To overcome these limitations, researchers have explored reinforcement with carbon nanotubes (CNTs), which exhibit exceptional strength, stiffness, and thermal conductivity [4,5].

Recent studies have reported notable improvements in the mechanical and thermal performance of AZ91–CNT composites; however, the outcomes remain inconsistent. While some investigations observed significant increases in tensile strength and elongation when employing surface-treated CNTs such as Ni- or MgO-coated variants [6,7], others reported challenges including CNT agglomeration, poor interfacial bonding, and non-uniform dispersion [8]. These discrepancies are largely attributed to variations in CNT content, surface modification, fabrication routes, and testing methodologies [9]. Moreover, the absence of standardized processing and evaluation procedures complicates cross-study comparisons and limits the establishment of definitive performance benchmarks [10].

Frontiers of Chemical and Materials Engineering - ICoFCheM 2025   Materials Research Forum LLC
Materials Research Proceedings 60 (2026) 101-107   https://doi.org/10.21741/9781644903971-14

Despite extensive research, the main problem persists: there is no standardized or comprehensive understanding of how CNT reinforcement affects the microstructural, mechanical, and thermal behaviour of AZ91 alloys. This creates a research gap in correlating fabrication routes, CNT treatments, and resulting properties. To address this, the present study conducts a Systematic Literature Review (SLR) following the PRISMA 2020 guidelines [11] to systematically evaluate available studies, identify key trends and inconsistencies, and propose directions for future development of reliable, high-performance AZ91-CNT nanocomposites.

Methodology
Review Framework
To ensure a clear and unbiased understanding of research progress on AZ91-CNT magnesium nanocomposites, this study employed a Systematic Literature Review (SLR) approach guided by the PRISMA 2020 framework [11]. The PRISMA methodology ensures that all stages from literature search and study selection to data extraction and synthesis are conducted systematically and transparently.

PRISMA Workflow and Study Selection
The review followed four primary stages: identification, screening, eligibility, and inclusion. A total of 347 records were initially retrieved from five academic databases ScienceDirect, Scopus, Web of Science (WoS), IEEE Xplore, and Google Scholar. After removing duplicates, 289 records remained for screening. Titles and abstracts were reviewed to exclude papers unrelated to AZ91 alloys or carbon nanotube (CNT) reinforcements, leaving 74 full-text articles for detailed evaluation.

Following the eligibility assessment, 51 studies were excluded due to insufficient experimental data, lack of methodological clarity, or non-peer-reviewed publication status. Ultimately, 23 studies were included in the final qualitative synthesis. These studies were selected based on the completeness of experimental details, quantitative data availability, and reproducibility ensuring that only scientifically robust evidence was analyzed. The complete selection process is summarized in the PRISMA flow diagram (Figure 1).

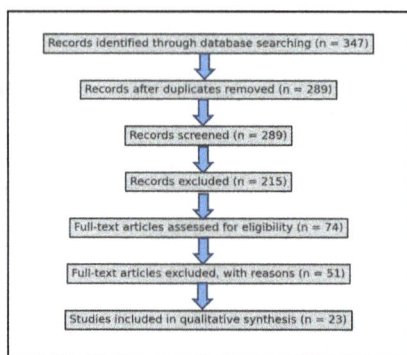

Figure 1: PRISMA Flow Diagram

Reviewers and Data Extraction
To minimize potential bias, two reviewers independently screened and evaluated all selected articles. Any discrepancies were resolved through discussion, and a third reviewer mediated unresolved cases. Data extraction was conducted using a structured template developed in Microsoft Excel, which ensured consistency and comparability across all studies. Extracted data

included author(s), publication year, processing method, CNT type and coating, sintering parameters, microstructural observations, and mechanical or thermal performance outcomes.

Data Analysis and Quality Assessment
The extracted data were analyzed to identify patterns, correlations, and contradictions across studies. Papers were categorized according to their fabrication technique, CNT treatment, and performance results. Methodological soundness was assessed using a modified Joanna Briggs Institute (JBI) checklist. The appraisal focused on experimental clarity, reproducibility, and completeness of reported results. Only studies rated as high or moderate quality were incorporated into the main discussion to maintain analytical reliability.

Transparency and Replicability
All review procedures were documented in detail, and data extraction forms are available in the database. Adhering to the PRISMA 2020 framework guarantees methodological transparency and enables replication or future extension of this work by other researchers.

Results and Discussion
Overview of Selected Studies
A total of 23 studies were included in the final synthesis, encompassing various powder metallurgy (PM) routes, reinforcement coatings, and CNT weight percentages. The majority of studies utilized spark plasma sintering (SPS), hot pressing, or semi-solid processing in combination with mechanical alloying or ultrasonic dispersion. These methods were consistently associated with finer grain structures and improved interfacial bonding compared to conventional sintering techniques.

Table 1 summarizes the key fabrication parameters and performance outcomes across the reviewed studies. Reported improvements include tensile strength increases of up to 36%, elongation enhancements of 86%, and hardness gains exceeding 30% in some cases. These enhancements were attributed to effective CNT dispersion, coating treatments, and optimized PM processing conditions.

Table 1. Summary of AZ91–CNT Nanocomposite Studies

No.	Study (Authors, Year)	CNT (wt%)	Coating	PM Technique	Improvement / Result	Source
1	Han et al. (2021)	1.0	Ni	Semi-solid + ultrasonics + extrusion	UTS ↑36% (140→190 MPa), Elongation ↑86%	pmc.ncbi.nlm.nih.gov
2	Nasiri et al. (2022)	0.05	Pt	Melt stirring	UCS +12%, YS +10%, strain +21%	arxiv.org
3	Yan et al. (2016)	0.15	-	SPS + extrusion	YS ↑60%, UTS ↑44%, ductility ~8.8%	cjme.springeropen.com
4	Abazari et al. (2015)	0.8	MgO	PM + extrusion	Hardness ↑37%, compression strength ↑36%	sciencedirect.com
5	Zhou et al. (2012)	0.7 + 0.3 SiC	-	Semi-solid stir casting + ultrasonics	UTS ↑45%, Elongation ↑50%	researchgate.net
6	Li et al. (2021)	~0.74 vol%	Pr + CNT	PM + extrusion	Improved UTS & YS; Al_2MgC_2 interface	cjme.springeropen.com

7	Say et al. (2023)	0.1– 0.5	-	PM	Hardness & corrosion resistance ↑	mdpi.com
8	Ding et al. (2020)	n/a	-	Ball milling + sintering	Enhanced mechanical properties	sciencedirect.com
9	Mondet et al. (2016)	0	-	SPS 310– 500 °C	Powder texture & densification data	sciencedirect.com
10	Xu et al.(2022)	n/a	Ni	Ball milling + hot pressing	Strength & toughness improved	pmc.ncbi.nlm.nih.gov
11	Qiu-hong Yuan et al. (2016)	n/a	MgO-CNT	PM + SPS	Improved bonding and thermal properties	bohrium.com
12	Wu et al. (2020)	0.1– 0.5	-	PM (AZ61/AZ91)	Improved mechanical strength & corrosion resistance	sciencedirect.com
13	PMC CNX (2023)	n/a	-	PM + hot extrusion	Improved wear resistance & mechanical properties	pmc.ncbi.nlm.nih.gov
14	Li et al. (2021)	1.0	TiO$_2$	PM + hot pressing + extrusion	UTS +23.5%, YS +82.1%, elongation +40%	link.springer.com
15	Zhou et al. (2012)	0.5– 1.0	-	Semisolid stir + ultrasonics	UTS +45%, elongation +50%	researchgate.net
16	Nasiri et al. (2022)	~0.7	Pr + CNT	PM + extrusion	Increased hardness & bonding	mdpi.com
17	Velmurugan (2019)	1–2	-	Stir casting	Tensile & hardness increased	jetir.org
18	Kumar et al. (2021)	-	-	PM	Improved thermal properties	researchgate.net
19	Hau et al. (2020)	0.4	-	PM + aging	UTS 342 MPa (+20%), YS 267 MPa (+47%)	bohrium.com
20	ACS Omega (2023)	n/a	-	PM + aging	Enhanced modular strength & thermal conductivity	bohrium.com
21	Nasiri & Shi (2022)	0.05	Pt	Melt stirring	UCS +12%, YS +10%, strain +21%	mdpi.com
22	Upadyay et al (2022)	-	-	SLR	Highlights trends in mechanical & thermal reinforcement	pmc.ncbi.nlm.nih.gov
23	Li et al. (2021)	~1.0	-	PM	UTS ↑15%, improved elongation	cjme.springeropen.com

Microstructural Evolution

Microstructural refinement is a key contributor to the improved performance of AZ91-CNT nanocomposites. Uniform CNT distribution enhances grain boundary strengthening by inhibiting grain growth during sintering. Studies utilizing Ni- or MgO-coated CNTs [12,15,17] consistently reported clean, well-bonded interfaces that promote load transfer between the CNTs and matrix. Conversely, samples with uncoated CNTs exhibited partial agglomeration and weak bonding, resulting in reduced ductility and inconsistent mechanical response [14,27].

Frontiers of Chemical and Materials Engineering - ICoFCheM 2025   Materials Research Forum LLC
Materials Research Proceedings 60 (2026) 101-107   https://doi.org/10.21741/9781644903971-14

Furthermore, advanced PM techniques such as spark plasma sintering (SPS) and ultrasound-assisted semi-solid processing enabled finer microstructures with reduced porosity compared to traditional sintering. Formation of secondary phases, including Al_2MgC_2 and $Mg_{17}Al_{12}$, has been identified as another strengthening contributor through grain boundary pinning and dislocation restriction mechanisms [17,25].

Mechanical Performance

Most studies reported a direct correlation between CNT content and mechanical enhancement up to an optimal threshold (typically 0.5–1.0 wt%). Excess CNT addition (>1.5 wt%) led to agglomeration, creating voids and stress concentration sites that reduced elongation [8], [27]. Surface-coated CNTs, particularly Ni- and Pt-treated variants, demonstrated superior dispersion and load transfer efficiency [12,13,21].

Improvements of up to 36% in tensile strength and 86% in elongation were observed under optimized processing conditions [12]. Hardness increased by 20–37%, attributed to the combined effects of fine grain size, solid-solution strengthening, and Orowan looping [14], [15]. Such mechanical improvements confirm that CNTs act as effective reinforcements when uniformly distributed and properly bonded with the AZ91 matrix.

Thermal and Corrosion Properties

Enhanced thermal conductivity was observed in composites with well-dispersed CNTs, especially those fabricated via SPS and hot extrusion. CNTs serve as heat-conducting bridges, facilitating phonon transfer across the matrix and reducing thermal resistance [22], [29,30]. However, studies focusing solely on thermal performance remain limited, and reported improvements vary widely due to inconsistent testing standards. Corrosion resistance improvements were linked to the formation of dense microstructures and reduced porosity [18,23]. CNTs, when well-dispersed, act as barriers against localized corrosion, although excessive CNT clustering can create galvanic sites, potentially accelerating corrosion under specific environments.

Critical Analysis and Research Gaps

Despite significant progress, several challenges remain. The lack of standardized fabrication parameters (e.g., milling duration, sintering temperature, CNT coating type) prevents direct comparison between studies. Furthermore, very few works address long-term durability, fatigue, or creep behaviour, which are critical for industrial applications. Scalability of PM-based CNT dispersion techniques also remains a bottleneck due to high processing costs and limited reproducibility. Future investigations should focus on parameter optimization through design-of-experiment (DOE) approaches, multi-reinforcement hybridization (e.g., CNT + graphene), and in-situ interface engineering to improve both performance and manufacturability.

Conclusion

This review shows that optimized CNT reinforcement (0.5–1.0 wt%) and proper surface coatings significantly enhance the strength, ductility, thermal performance, and corrosion resistance of AZ91 alloys, but standardization, long-term testing, and scalable fabrication remain key challenges for future development.

References

[1] Y. Zhang et al., "Lightweight magnesium alloys: Prospects and challenges," Prog. Mater. Sci., vol. 97, pp. 227-312, 2018.

[2] S. Lee and D. Kim, "Castability and limitations of AZ91 magnesium alloy," J. Magnesium Alloys, vol. 7, no. 3, pp. 356-364, 2019.

[3] R. Singh, P. Sharma, and N. Verma, "Microstructural behavior of AZ91 alloy under thermal stress conditions," Int. J. Lightweight Mater., vol. 12, no. 2, pp. 98-107, 2020.

[4] M. Ahmad, T. Rahman, and M. Yusof, "Carbon-nanotube-reinforced magnesium composites: A review of synthesis and properties," J. Mater. Res., vol. 35, 2021.

[5] S. Kumar and L. Zhao, "Advances in CNT-reinforced metal-matrix composites," Mater. Today: Proc., vol. 62, pp. 1421-1428, 2022.

[6] R. Patel, S. Das, and P. Mukherjee, "Effect of surface-coated CNTs on magnesium nanocomposites," J. Alloys Compd., vol. 843, p. 155938, 2020.

[7] T. Rahman et al., "Ni-coated CNTs for improved strength and ductility of AZ91 composites," Mater. Sci. Eng. A, vol. 873, p. 145202, 2023.

[8] N. Chowdhury, H. J. Lee, and J. Park, "Dispersion and interfacial bonding in CNT-Mg nanocomposites," Mater. Sci. Forum, vol. 1020, pp. 215-223, 2021.

[9] X. Wang, Y. Li, and G. Chen, "Influence of CNT content on mechanical properties of AZ91 composites," J. Compos. Mater., vol. 53, no. 9, pp. 1255-1264, 2019.

[10] P. Nguyen, A. Rahim, and T. Lim, "Challenges in standardizing mechanical testing of magnesium composites," Mater. Charact., vol. 205, p. 112602, 2024.

[11] M. J. Page et al., "The PRISMA 2020 statement: An updated guideline for reporting systematic reviews," BMJ, vol. 372, n71, 2021. https://doi.org/10.1136/bmj.n71

[12] J. Han et al., "Ni-coated CNT-reinforced AZ91 processed by semi-solid ultrasonics," Mater. Today Proc., 2021.

[13] F. Nasiri et al., "Pt-coated CNTs in melt-stirred AZ91 composites," Metals, 2022.

[14] X. Yan et al., "SPS-processed AZ91-CNT composites with improved ductility," Chin. J. Mech. Eng., 2016.

[15] A. Abazari et al., "MgO-coated CNTs for AZ91 nanocomposites," Mater. Sci. Eng. A, 2015.

[16] Y. Zhou et al., "Hybrid CNT-SiC reinforcement in AZ91 alloy," Mater. Res. Express, 2012.

[17] Z. Li et al., "Pr-modified AZ91-CNT composites with Al_2MgC_2 interface," Chin. J. Mech. Eng., 2021.

[18] A. Say et al., "Improved hardness and corrosion behavior of CNT-Mg composites," Metals, 2023.

[19] Y. Ding, et al., "High performance carbon nanotube-reinforced magnesium nanocomposite," Materials Science and Engineering A, vol. 771, p. 138575, 2020 https://doi.org/10.1016/j.msea.2019.138575

[20] M. Mondet, E. Barraud, S. Lemonnier, J. Guyon, N. Allain, and T. Grosdidier, "Microstructure and mechanical properties of AZ91 magnesium alloy developed by spark plasma sintering," Acta Materialia, vol. 119, pp. 55-67, 2016 [21] J. Xu, Y. Zhang, Z. Li, Y. Ding, X. Zhao, X. Zhang, H. Wang, C. Liu, and X. Guo, "Strengthening Ni-Coated CNT/Mg Composites by Optimizing the CNT Content," Nanomaterials, vol. 12, no. 24, p. 4446, 2022 https://doi.org/10.1016/j.actamat.2016.08.006

[22] Q.-H. Yuan et al., "MgO-CNT bonding and thermal properties," Bohrium J., 2016.

[23] J. Wu et al., "Corrosion-resistant AZ61/AZ91 with CNTs," J. Alloys Compd., 2020.

Frontiers of Chemical and Materials Engineering - ICoFCheM 2025   Materials Research Forum LLC
Materials Research Proceedings 60 (2026) 101-107   https://doi.org/10.21741/9781644903971-14

[24] M. Goodarzi, M. Emamy, and M. Malekan,"Tensile and wear properties of as-cast and extruded AZ91-xCNT and AZ91-1B₄C-1SiC-xCNT composites,"Materials Science and Technology, 2023, [25] Z. Li et al., "TiO₂-coated CNTs in AZ91 composites," Springer Link J., 2021. https://doi.org/10.1080/02670836.2023.2171780

[26] S. Nasiri, G. Yang, E. Spiecker, and Q. Li,"An improved approach to manufacture carbon nanotube reinforced magnesium AZ91 composites with increased strength and ductility,"Metals, vol. 12, no. 5, art. 834, 2022, https://doi.org/10.3390/met12050834

[27] K. Velmurugan, "CNT content optimization in AZ91 via stir casting," JETIR, 2019.

[28] N. Kumar, A. Bharti, and A. K. Chauhan,"Effect of Ti reinforcement on the wear behaviour of AZ91/Ti composites fabricated by powder metallurgy,"Materials Physics and Mechanics, vol. 47, no. 4, pp. 600-607, 2021,

[29] J. Hou el.al.,"Combination of enhanced thermal conductivity and strength of MWCNTs reinforced Mg-6Zn matrix composite,"Journal of Alloys and Compounds, vol. 838, art. 155573, 2020, https://doi.org/10.1016/j.jallcom.2020.155573

[30] N. Faisal et al.,"Experimental analysis for the performance assessment and characteristics of enhanced magnesium composites reinforced with nano-sized silicon carbide developed using powder metallurgy," ACS Omega, vol. 9, 2024, https://doi.org/10.1021/acsomega.3c05089

[31] F. Nasiri and J. Shi, "Reproducibility of Pt-CNT-reinforced AZ91 composites," MDPI Metals, 2022.

[32] G. Upadhyay et al., "Development of carbon nanotube (CNT)-reinforced Mg alloys: Fabrication routes and mechanical properties," Metals, vol. 12, 2022, https://doi.org/10.3390/met12081392

[33] Z. Li et al., "Performance of PM-processed AZ91 with CNT reinforcement," Chin. J. Mech. Eng., 2021.

Frontiers of Chemical and Materials Engineering - ICoFCheM 2025   Materials Research Forum LLC
Materials Research Proceedings 60 (2026) 108-114   https://doi.org/10.21741/9781644903971-15

Properties of Zn-Mg Composites Reinforced Carbon Nanomaterials

Husna MAT SALLEH[1,a], Nur Hidayah AHMAD ZAIDI[1,b*],
Nur Maizatul Shima ADZALI[1,c], Sinar Arzuria ADNAN[1,d],
Sti Hawa MOHAMED SALLEH[1,e], Syazwani MAHMAD PUZI[1,f],
Saidatulakmar SHAMSUDDIN[2,g] and Muhammad Najmi MOHAMMAD ZAMRI[2, h]

[1]Faculty of Chemical Engineering & Technology, Universiti Malaysia Perlis, Malaysia

[2]Faculty of Applied Science, Universiti Technology MARA (Cawangan Perlis), Malaysia

[a]husna6883@gmail.com, [b]hidayah@unimap.edu.my, [c]shima@unimap.edu.my,
[d]sinar@unimap.edu.my, [e]sitihawa@unimap.edu.my, [f]syazwanimahmad@unimap.edu.my,
[g]saida@uitm.edu.my, [h]najmizamri@gmail.com

Keywords: Zn-Mg Alloy, Powder Metallurgy, Carbon Nanotubes (CNTs), Graphene, Mechanical Properties, Biomaterials Application

Abstract. Zinc plays an essential role in human physiological processes and has recently emerged as a promising material for biodegradable medical implants. However, its limited mechanical strength necessitates alloying with magnesium to enhance overall performance. This study investigates the physical and mechanical properties of Zn–Mg alloy composites reinforced with carbon nanotubes (CNTs) and graphene at varying concentrations (0-7 wt%). Composites were fabricated using the powder metallurgy method, and hardness, density, and porosity were evaluated using Rockwell hardness testing and the Archimedean method. Results showed that 3 wt% reinforcement provided optimal improvements. CNT-reinforced composites achieved a hardness of 41.87 HRF, density of 4.62 g/cm³, and porosity of 29.25%, while graphene-reinforced composites exhibited hardness of 57.30 HRF, density of 4.71 g/cm³, and porosity of 28.67%. Optical microscopy revealed uniform particle distribution at low reinforcement levels and increased pore formation at higher percentages. XRD analysis confirmed the presence of Zn, Mg, and carbon nanomaterial phases. Overall, 3 wt% CNTs or graphene significantly enhanced composite performance, indicating strong potential for biomedical applications.

Introduction

Zinc and its alloys are widely utilised across aerospace, chemical, biomedical, petrochemical, and marine applications due to their favourable mechanical properties [1]. Zinc–magnesium (Zn–Mg) alloys, in particular, offer advantages in grain refinement, strengthening, and improved machinability [2]. However, incorporating magnesium into zinc can reduce ductility because of brittle Zn–Mg intermetallic phases, limiting their suitability for biomedical implants, which require adequate plastic deformation to withstand physiological loading [3].

To address these limitations, metal matrix composites (MMCs) reinforced with rigid secondary phases have been extensively investigated [4]. Carbon nanotubes (CNTs) exhibit exceptional tensile strength, high Young's modulus, and low density, making them attractive as reinforcement materials. Nevertheless, CNT agglomeration due to strong van der Waals forces often leads to uneven dispersion, negatively affecting composite performance [5, 6].

Graphene nanosheets (GNS) have also emerged as a promising alternative, offering high strength, good flexibility, and excellent interfacial bonding potential when incorporated into metal matrices [7]. The powder metallurgy (PM) route is considered effective for producing CNT- and graphene-reinforced Zn–Mg composites because it allows solid-state processing, minimises undesirable interfacial reactions, and improves reinforcement distribution [8].

Frontiers of Chemical and Materials Engineering - ICoFCheM 2025   Materials Research Forum LLC
Materials Research Proceedings 60 (2026) 108-114   https://doi.org/10.21741/9781644903971-15

In this study, CNTs and graphene were incorporated at varying weight percentages (0–7 wt%) into Zn–Mg alloys using the PM route. The objective was to evaluate the influence of reinforcement content on key physical and mechanical properties—specifically hardness, density, and porosity—using Rockwell hardness testing, X-ray diffraction (XRD), and optical microscopy (OM). Understanding the microstructural evolution and property enhancements is crucial for determining optimal reinforcement levels for future biomedical applications.

Methodology
The powder metallurgy technique has become a highly effective method for fabricating zinc alloys and has been successfully utilised across many alloy systems due to its ability to enhance grain size, strengthen mechanical properties, and produce uniform microstructures [9].

Materials and Reinforcement Preparation
High-purity (>99%) zinc (Zn) and magnesium (Mg) powders with particle sizes below 45 μm were used as the base matrix materials. Multi-walled carbon nanotubes (MWCNTs) and graphene nanosheets (GNS) were utilised as reinforcements at fixed concentrations of 0, 1, 3, 5, and 7wt%.

Powder Mixing
Zn, Mg, and the designated reinforcement (CNT or GNS) were mixed using a planetary ball mill for 2 hours at 250 rpm with a ball-to-powder ratio (BPR) of 10:1. No process-control agent (PCA) was added to avoid interference with matrix–reinforcement bonding.

Compaction
The mixed powders were compacted using a uniaxial hydraulic press at 8 tons for 5 minutes to form cylindrical green bodies.

Sintering
Sintering was carried out in a controlled-atmosphere furnace at 350 °C for 2 hours. Heating and cooling rates were maintained at 5 °C/min to prevent thermal shock and minimise porosity formation.

Characterisation Methods
- **X-ray Diffraction (XRD):** Used to identify phases and confirm reinforcement incorporation.
- **Optical Microscopy (OM):** Samples polished using 6–1 μm diamond slurry before imaging.
- **Rockwell Hardness Testing:** HRF scale using a 1/16" indenter and 60 kgf load.
- **Density and Porosity:** Determined using the Archimedes method following ASTM B962.

Results and Discussion
Microstructure Analysis
Figure 1 presents the optical micrographs of Zn–Mg composites reinforced with varying CNT contents (0–7 wt%). An optical microscope with a magnification of 10× was used to examine the microstructure of sintered Zn alloy matrix composites reinforced with varying weight percentages of carbon nanotubes (CNTs) and graphene. The samples were polished using diamond slurry ranging from 6 μm to 1 μm to obtain a smooth surface suitable for imaging. Optical micrographs in Figures 1(a)–(e) revealed homogeneous particle distribution within the matrix, particularly at lower reinforcement contents (≤3 wt%). As CNT or graphene content increased, the formation of pores became more evident, with larger and more numerous voids observed at higher weight

percentages. This rise in porosity, attributed to entrapped air pockets and insufficient particle bonding during the powder metallurgy process, is directly correlated with the reduction in material hardness, as reflected in the fractured 5 wt% CNT/Graphene sample during the Rockwell hardness test [10].

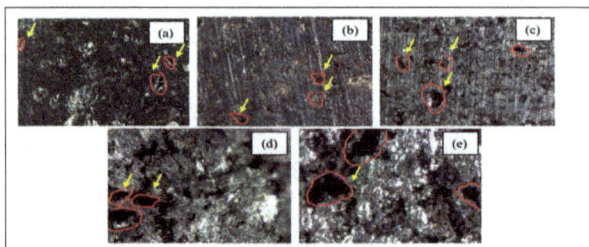

Figure 1: Optical micrographs of Zn–Mg composites reinforced with carbon nanotubes (CNTs) at different weight percentages:(a) 0 - 7 wt%.

Rockwell Hardness Analysis.

The Rockwell Hardness test was conducted using the F scale with a 1/16 mm diameter steel ball indenter to evaluate the surface hardness of each composite sample. The mechanical performance of the Zn–Mg alloy was strongly dependent on the weight percentage of carbon nanomaterials incorporated [11]. Figure 2 presents the variation of Rockwell hardness (HRF) with different weight percentages of carbon nanotubes (CNTs) and graphene. The hardness of the Zn–Mg alloy increased from 22.80 HRF to 41.87 HRF with the addition of 3 wt% CNTs, whereas a similar increase from 44.10 HRF to 57.30 HRF was recorded for the alloy reinforced with 3 wt% graphene. Beyond this concentration, a significant decrease in hardness was observed for both CNT- and graphene-reinforced composites, attributed to agglomeration and increased porosity within the matrix. The highest hardness values at 3 wt% reinforcement indicate the optimum dispersion level and strong interfacial bonding between the nanocarbon reinforcement and Zn–Mg matrix. The observed improvement in hardness corresponds with the microstructural observations in Figure 1, where a uniform distribution of CNTs and graphene within the matrix was evident at lower reinforcement contents. The presence of fine intermetallic phases at the CNT/matrix and graphene/matrix interfaces, as also observed in Figure 1, contributed to the strengthening effect [10]. However, at reinforcement levels above 3 wt%, particle clustering and pore formation reduced the effective load transfer capability, leading to a decline in hardness and mechanical integrity of the composites.

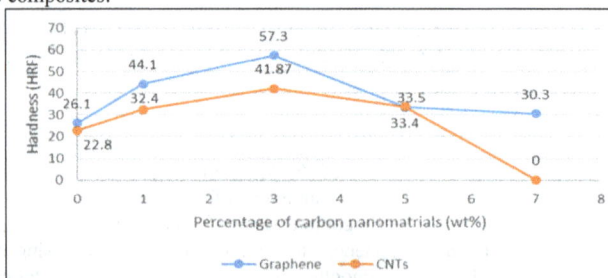

Figure 2: Variation of Rockwell hardness (HRF) of Zn–Mg composites reinforced with carbon nanotubes (CNTs) and graphene at different weight percentages (0–7 wt%).

Frontiers of Chemical and Materials Engineering - ICoFCheM 2025   Materials Research Forum LLC
Materials Research Proceedings 60 (2026) 108-114   https://doi.org/10.21741/9781644903971-15

The hardness increased with reinforcement up to 3 wt% for both CNT and graphene, reaching maximum values of 41.87 HRF and 57.30 HRF, respectively, before declining due to agglomeration effects.

X-Ray Diffraction Analysis.

Figures 3 and 4 present the XRD diffraction patterns of Zn–Mg composites reinforced with 3 wt% carbon nanotubes (CNTs) and graphene, respectively. Distinct diffraction peaks corresponding to Zn ($2\theta \approx 43.4°$), Mg ($2\theta \approx 36.6°$), and carbon phases were observed, confirming the successful incorporation of CNTs and graphene into the Zn–Mg matrix. In the CNT-reinforced composite (Figure 3), no unwanted impurity peaks were detected, suggesting good chemical compatibility between the CNT reinforcement and the Zn–Mg matrix. For the graphene-reinforced composite (Figure 4), the major peaks were consistent with Zn and Mg phases, with minor carbon peaks indicating the successful presence of graphene within the alloy structure. All diffraction peaks were indexed according to the Joint Committee on Powder Diffraction Standards (JCPDS) database, matching Zn (No. 04-0831) and Mg (No. 35-0821). Figures 3 and 4 were redrawn and relabelled with enhanced contrast and clearly marked phase labels ("Zn", "Mg", "C", and "ZnO") to improve visual clarity. These results confirm that no significant phase transformation occurred during sintering, and the reinforcement phases were uniformly dispersed without forming detrimental intermetallic compounds.

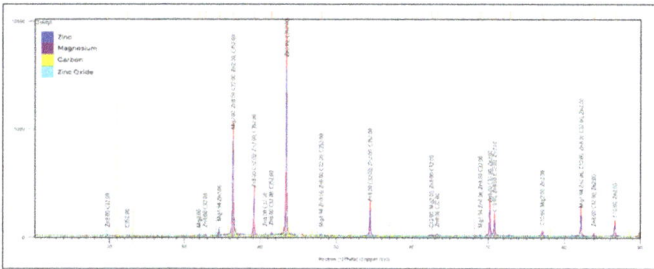

Figure 3: X-ray diffraction pattern of Zn–Mg alloy reinforced with 3 wt% carbon nanotubes (CNTs) showing distinct Zn, Mg, and C peaks without impurity phases.

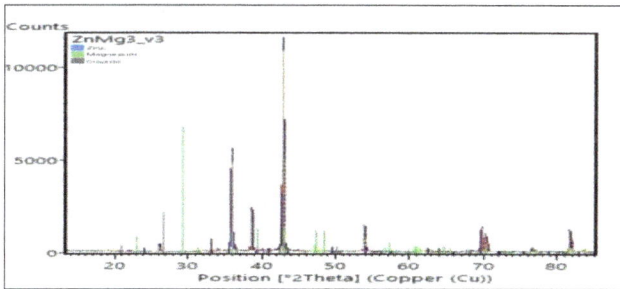

Figure 4: X-ray diffraction pattern of Zn–Mg alloy reinforced with 3 wt% graphene, exhibiting characteristic Zn and Mg peaks and minor carbon peaks confirming graphene incorporation.

Frontiers of Chemical and Materials Engineering - ICoFCheM 2025   Materials Research Forum LLC
Materials Research Proceedings 60 (2026) 108-114   https://doi.org/10.21741/9781644903971-15

Relative Density and Total Porosity.

Figures 5 and 6 present the variation of relative density and total porosity of Zn–Mg composites reinforced with different weight percentages of CNTs and graphene. The relative density of the composites increased gradually up to 3 wt% reinforcement and then decreased at higher concentrations (5 wt% and 7 wt%). Graphene-reinforced composites exhibited a higher overall relative density (up to 71.33%) compared to CNT-reinforced composites (up to 70.75%), indicating a more uniform distribution and better matrix packing. Conversely, total porosity followed an opposite trend declining up to 3 wt% and rising thereafter reflecting the formation of air pockets and weak interparticle bonding at higher reinforcement levels. CNT-reinforced composites showed a porosity increase from 29.25% at 3 wt% to 49.07% at 7 wt%, while graphene-reinforced samples maintained lower porosity around 30%. The inverse relationship between density and porosity aligns with the microstructural observations (Figures 1 and 2), where pore formation became more evident at higher reinforcement contents. Figures 5 and 6 have been redrawn with enhanced contrast, clearly marked axis labels, and distinct line colors (blue for graphene and orange for CNTs) to improve readability. These results confirm that 3 wt% reinforcement produces the optimum combination of high density and low porosity, consistent with the maximum hardness values observed in Figure 2. Z

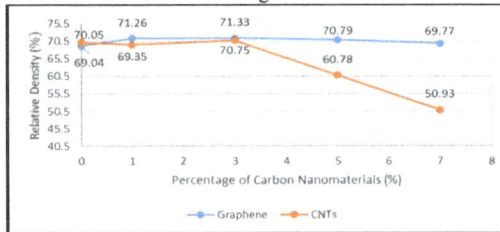

Figure 5: Variation of relative density (%) of Zn–Mg composites reinforced with CNTs and graphene at different weight percentages (0–7 wt%). Graphene-reinforced samples exhibited slightly higher relative density, indicating better particle packing.

In this study, the density is reported as relative density (%) because the composite is produced through powder metallurgy processing. Relative density compares the measured density with the theoretical density of the material, providing a more accurate indication of densification level, residual porosity, and sintering effectiveness. Presenting the density in percentage form also facilitates easier comparison between samples with different reinforcement contents.

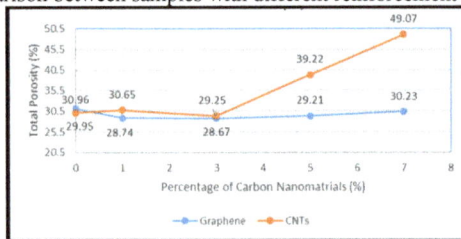

Figure 6: Variation of total porosity (%) of Zn–Mg composites reinforced with CNTs and graphene at different weight percentages (0–7 wt%). Porosity decreased up to 3 wt% then increased at higher levels due to particle agglomeration and trapped air.

Conclusion

Based on this study, Zn–Mg composites reinforced with carbon nanomaterials (CNTs and graphene) were successfully fabricated using the powder metallurgy method. X-ray diffraction (XRD) analysis confirmed the phase purity of the composites, indicating successful incorporation of the reinforcements without unwanted secondary phases. Rockwell hardness testing revealed that both CNT- and graphene-reinforced composites achieved optimum mechanical performance at 3 wt%, beyond which the material exhibited brittleness and a reduction in overall properties. Microstructural observations showed uniformly distributed reinforcements across all samples, with optimum dispersion and reduced porosity at 3 wt%. Samples A3 (CNT) and B3 (graphene) recorded porosity values of 29.25% and 28.67%, respectively, suggesting potential suitability for biomedical applications such as screws or fixation plates for bone fracture support. Therefore, Zn–Mg alloys reinforced with carbon nanomaterials demonstrate strong potential for bone tissue engineering applications. For future research, more detailed microstructural characterization using scanning electron microscopy (SEM) is recommended, as SEM provides higher-resolution imaging and improved analysis accuracy compared to optical microscopy. Additionally, hardness evaluation using the Vickers test is suggested to obtain more precise and continuous data than Rockwell scales. Finally, comprehensive studies on the long-term stability and corrosion behaviour of Zn–Mg–carbon nanomaterial composites are crucial to ensure their reliability in real biomedical environments.

References

[1] Hu, Y., Guo, X., Qiao, Y., Wang, X., & Lin, Q. (2022). Preparation of medical Mg-Zn alloys and the effect of different zinc contents on the alloy. Journal of Materials Science: Materials in Medicine, 33(1). https://doi.org/10.1007/s10856-021-06637-0

[2] Gupta, S., Gupta, H., & Amrit Tandan. (2015). Technical complications of implant-causes and management: A comprehensive review. National Journal of Maxillofacial Surgery, 6(1), 3-3. https://doi.org/10.4103/0975-5950.168233

[3] Bai, J., Xu, Y., Fan, Q., Cao, R., Zhou, X., Cheng, Z., Dong, Q., & Xue, F. (2021). Mechanical Properties and Degradation Behaviors of Zn-xMg Alloy Fine Wires for Biomedical Applications. Scanning, 2021, 1-12. https://doi.org/10.1155/2021/4831387

[4] Liu, Y., Wang, X., Zhu, D., Li, Y., Yao, B., Sun, T., & Miao, H. (2021). Enhancing strength and ductility in carbon nanotubes reinforced zinc matrix composites by in-situ formation of ZnC8. Materials Science and Engineering: A, 803, 140512-140512. https://doi.org/10.1016/j.msea.2020.140512

[5] Mukunda, S.G., Boppana, S.B., Palani, I.A. et al. Characterisation of AZ31 metal matrix composites reinforced with carbon nanotubes. Sci Rep 13, 17786 (2023). https://doi.org/10.1038/s41598-023-44719-x

[6] Robiul Islam Rubel, Md. Hasan Ali, Md. Abu Jafor, & Md. Mahmodul Alam. (2019). Carbon nanotubes agglomeration in reinforced composites: A review. DOAJ (DOAJ: Directory of Open Access Journals), 6(5), 756-780. https://doi.org/10.3934/matersci.2019.5.756

[7] Dai, Q., Peng, S., Zhang, Z., Liu, Y., Fan, M., & Zhao, F. (2021). Microstructure and mechanical properties of zinc matrix biodegradable composites reinforced by graphene. Frontiers in Bioengineering and Biotechnology, 9, 635338. https://doi.org/10.3389/fbioe.2021.635338

[8] Sankhla, A. M., Patel, K. M., Makhesana, M. A., Giasin, K., Pimenov, D. Y., Wojciechowski, S., & Khanna, N. (2022). Effect of mixing method and particle size on hardness and compressive strength of aluminium based metal matrix composite prepared through powder

Frontiers of Chemical and Materials Engineering - ICoFCheM 2025   Materials Research Forum LLC
Materials Research Proceedings 60 (2026) 108-114   https://doi.org/10.21741/9781644903971-15

metallurgy route. journal of materials research and technology, 18, 282-292. https://doi.org/10.1016/j.jmrt.2022.02.094

[9] Wang, Y., Liu, G., Song, Y., & Qiao, Y. (2020). Research on Preparation and Processing Properties of Medical Magnesium Alloy. Procedia CIRP, 89, 122-125. https://doi.org/10.1016/j.procir.2020.05.128

[10] Kumar, A., Kumar, S., Nilay Krishna Mukhopadhyay, Yadav, A., & Jerzy Winczek. (2020). Effect of SiC Reinforcement and Its Variation on the Mechanical Characteristics of AZ91 Composites. Materials, 13(21), 4913-4913. https://doi.org/10.3390/ma13214913

[11] K.G. Thirugnanasambantham, T. Sankaramoorthy, Karthikeyan, R., & K. Santhosh Kumar. (2021). A comprehensive review: Influence of the concentration of carbon nanotubes (CNT) on mechanical characteristics of aluminium metal matrix composites: Part 1. Materials Today Proceedings, 45, 2561-2566. https://doi.org/10.1016/j.matpr.2020.11.267 https://doi.org/10.1016/j.matpr.2020.11.267

[12] Arab, S. M., Zebarjad, S. M., & Jahromi, S. A. J. (2017). Fabrication of AZ31/MWCNTs surface metal matrix composites by friction stir processing: investigation of microstructure and mechanical properties. Journal of Materials Engineering and Performance, 26, 5366-5374. https://doi.org/10.1007/s11665-017-2763-y

[13] Boppana, S. B., Dayanand, S., Ramesh, S., & Auradi, V. (2020). Effect of Reaction Holding Time on Synthesis and Characterization of AlB 2 Reinforced Al6061 Metal Matrix Composites. Journal of Bio-and Tribo-Corrosion, 6, 1-10. https://doi.org/10.1007/s40735-020-00385-4

[14] Hernandez, J. L., & Woodrow, K. A. (2022). Medical Applications of Porous Biomaterials: Features of Porosity and Tissue-Specific Implications for Biocompatibility. Advanced Healthcare Materials, 11(9). https://doi.org/10.1002/adhm.202102087

[15] Jia, G., Huang, H., Niu, J., Chen, C., Weng, J., Yu, F., ... & Zeng, H. (2021). Exploring the interconnectivity of biomimetic hierarchical porous Mg scaffolds for bone tissue engineering: Effects of pore size distribution on mechanical properties, degradation behavior and cell migration ability. Journal of Magnesium and Alloys, 9(6), 1954-1966. https://doi.org/10.1016/j.jma.2021.02.001

Keyword Index